Diversity in Engineering

Managing the Workforce of the Future

NATIONAL ACADEMY PRESS
Washington, D.C.

NATIONAL ACADEMY PRESS • 2101 Constitution Avenue, N.W. • Washington, DC 20418

NOTICE: The project that is the subject of this report was approved by the Governing Board of the National Research Council, whose members are drawn from the councils of the National Academy of Sciences, the National Academy of Engineering, and the Institute of Medicine. The members of the committee responsible for the report were chosen for their special competences and with regard for appropriate balance.

This workshop was supported by Grant No. DE-FG02-00ER76080 between the National Academy of Sciences and the U.S. Department of Energy, with additional support from the Dow Chemical Company Foundation, the Motorola Foundation, the GE Fund, DuPont Engineering, and the National Academy of Engineering. Any opinions, findings, conclusions, or recommendations expressed in this publication are those of the author(s) and do not necessarily reflect the views of the organizations or agencies that provided support for the project.

International Standard Book Number 0-309-08429-6

Additional copies of this report are available from:

National Academy Press
2101 Constitution Avenue, N.W.
Lockbox 285
Washington, DC 20055
(800) 624-6242
(202) 334-3313 (in the Washington metropolitan area)
http://www.nap.edu

Printed in the United States of America
Copyright 2002 by the National Academy of Sciences. All rights reserved.

THE NATIONAL ACADEMIES

National Academy of Sciences
National Academy of Engineering
Institute of Medicine
National Research Council

The **National Academy of Sciences** is a private, nonprofit, self-perpetuating society of distinguished scholars engaged in scientific and engineering research, dedicated to the furtherance of science and technology and to their use for the general welfare. Upon the authority of the charter granted to it by the Congress in 1863, the Academy has a mandate that requires it to advise the federal government on scientific and technical matters. Dr. Bruce M. Alberts is president of the National Academy of Sciences.

The **National Academy of Engineering** was established in 1964, under the charter of the National Academy of Sciences, as a parallel organization of outstanding engineers. It is autonomous in its administration and in the selection of its members, sharing with the National Academy of Sciences the responsibility for advising the federal government. The National Academy of Engineering also sponsors engineering programs aimed at meeting national needs, encourages education and research, and recognizes the superior achievements of engineers. Dr. Wm. A. Wulf is president of the National Academy of Engineering.

The **Institute of Medicine** was established in 1970 by the National Academy of Sciences to secure the services of eminent members of appropriate professions in the examination of policy matters pertaining to the health of the public. The Institute acts under the responsibility given to the National Academy of Sciences by its congressional charter to be an adviser to the federal government and, upon its own initiative, to identify issues of medical care, research, and education. Dr. Kenneth I. Shine is president of the Institute of Medicine.

The **National Research Council** was organized by the National Academy of Sciences in 1916 to associate the broad community of science and technology with the Academy's purposes of furthering knowledge and advising the federal government. Functioning in accordance with general policies determined by the Academy, the Council has become the principal operating agency of both the National Academy of Sciences and the National Academy of Engineering in providing services to the government, the public, and the scientific and engineering communities. The Council is administered jointly by both Academies and the Institute of Medicine. Dr. Bruce M. Alberts and Dr. Wm. A. Wulf are chairman and vice chairman, respectively, of the National Research Council.

COMMITTEE ON DIVERSITY IN THE ENGINEERING WORKFORCE

Cordell Reed, *chair,* Senior Vice President (retired), Commonwealth Edison Company
Sandra Begay-Campbell, Senior Member of Technical Staff, Sandia National Laboratories
Suzanne Brainard, Director, Center for Workforce Development, University of Washington
Daryl Chubin, Senior Vice President, Policy and Research, National Action Council for Minorities in Engineering
Jose B. Cruz, Jr., Howard D. Winbigler Chair and Professor of Electrical Engineering, Ohio State University
William Friend, Executive Vice President and Director (retired), Bechtel Group, Inc.
Suzanne Jenniches, Vice President of Communications Systems, Northrop-Grumman Corporation
Cathy Lasser, Vice President B2B Initiatives, IBM Corporation
Benjamin Montoya, President and CEO (retired), Public Service Company of New Mexico
David M. Porter, Jr., Assistant Professor of Human Resources and Organizational Behavior, Anderson Graduate School of Management, University of California-Los Angeles
James West, Senior Research Scientist, Bell Laboratories

NAE Staff
Peggy Layne, NAE Fellow and Program Officer
Nathan Kahl, Senior Project Assistant
Katie Gramling, Research Assistant
Proctor Reid, Associate Director, Program Office
Carol R. Arenberg, Managing Editor

Preface

Proportionately fewer women and underrepresented minorities are found in the engineering profession than in the U.S. workforce in general and in all other scientific or technical fields. As Wm. A. Wulf, president of the National Academy of Engineering (NAE), has noted, for the United States to remain competitive in a global technological society, the country as a whole must take serious steps to ensure that we have a diverse, well trained, multicultural workforce. To further that goal, the NAE established the Program on Diversity in the Engineering Workforce, which leads the NAE's efforts to increase diversity in the U.S. engineering workforce, by bringing together stakeholders from industry, government, academia, and nonprofit organizations to share their knowledge, identify needs, and discuss ways to address those needs.

As first steps towards addressing this issue, the NAE convened the "Summit on Women in Engineering" in May 1999, developed the "Celebration of Women in Engineering" web site (*www.nae.edu/cwe*), and convened a workshop to develop a business case for diversity in September 1999. The Committee on Diversity in the Engineering Workforce was established in November 1999, followed by the Forum on Diversity in the Engineering Workforce in January 2000. The workshop, "Best Practices in Managing Diversity," is the latest in the NAE's ongoing efforts to increase diversity in engineering.

The workshop brought together leaders of corporations that have been recognized for outstanding diversity programs and members of the NAE Committee and the NAE Forum on Diversity in the Engineering Workforce. Many thanks to the members of the committee and forum for their assistance in identifying and recruiting workshop speakers and participants and facilitating and reporting on breakout discussions.

The papers in this volume represent the authors' views as presented at the workshop. The questions and answers and summaries of the breakout discussions were taken from a transcript of the workshop. The participants did not attempt to provide any formal conclusions or recommendations, focusing instead on collecting information and informing the discussion of the issues.

Contents

Executive Summary	1
Introduction *Cordell Reed*	6
The Importance of Diversity in Engineering *Wm. A. Wulf*	8

BEST PRACTICES IN MANAGING DIVERSITY:
A PANEL DISCUSSION

Introduction *Mary C. Mattis*	17
Diversity Practices at CH2M HILL *Daniel E. Arvizu*	20
Diversity Practices at Motorola *Iwona Turlik*	24
Diversity Practices at Consolidated Edison *Richard P. Cowie*	30
Question and Answer Session	33

SUMMARIES OF MORNING BREAKOUT SESSIONS

Recruitment, Retention, Advancement: What Works?	41

VOICES FROM THE FIELD: A PANEL DISCUSSION

National Society of Black Engineers 51
Michele Lezama
Society of Women Engineers 56
Shelley A.M. Wolff
Society of Hispanic Professional Engineers 59
Orlando A. Gutierrez
American Indian Science and Engineering Society 61
Sandra Begay-Campbell
Question and Answer Session 65

AFTERNOON KEYNOTE ADDRESS

Advancing Minorities in Science and Engineering Careers 71
Willie Pearson, Jr.

AFTERNOON PRESENTATIONS AND SUMMARIES OF BREAKOUT SESSIONS

Affirmative Action Backlash 79
Tyrone D. Taborn
A Case Study of the Texaco Lawsuit 85
Thomas S. Williamson, Jr.
Diversity in the Global Marketplace 92
Lisa Nungesser
Mentoring 99
Janet M. Graham and Sarah Ann Kerr

MORNING KEYNOTE ADDRESS

Implementing Change 105
Nicholas Donofrio

SUMMARIES OF FINAL BREAKOUT SESSIONS

Where Do We Go from Here? 117

CLOSING ADDRESS

The Business Case for Diversity 125
James J. Padilla

APPENDIXES

A: Workshop Participants 137

B: Biographical Sketches of Workshop Speakers 142

C: Corporate Benchmarks 149

Executive Summary

On October 29 and 30, 2001, the National Academy of Engineering (NAE) Committee on Diversity in the Engineering Workforce brought together representatives of corporations that have been recognized for their successful diversity programs and members of the NAE Forum on Diversity to participate in a workshop entitled "Best Practices in Managing Diversity." The purpose of the workshop was to identify and describe corporate programs that have successfully recruited, retained, and advanced women and underrepresented minorities in engineering careers and to discuss metrics by which to evaluate diversity programs. The workshop was focused primarily on personnel policies and programs for engineers employed in industry and consulting services.

In preparation for the workshop, the committee compiled a list of companies that employ significant numbers of engineers that have been recognized for their handling of diversity issues in the workplace. These organizations were listed in the *Fortune* top 50 companies for minorities or *Working Woman's* best companies for women, have been recognized by Catalyst or the Women in Engineering Programs and Advocates Network for their diversity programs, are represented on the board of directors of the National Action Council for Minorities in Engineering, or were recommended by a member of the committee. The committee collected available information about the diversity management practices used by these companies as benchmarks for the workshop participants. These practices are included as an appendix to these proceedings.

The format of the workshop, which included plenary presentations followed by small group ("breakout") discussions, was designed to stimulate interaction among participants, as well as with the speakers. At the conclusion of the workshop, the participants were challenged to implement the best practices identified at the workshop in their respective organizations.

NAE president William A. Wulf gave the keynote address. In it, he stressed the importance of diversity and creativity in solving engineering problems. He emphasized the advantages of engineering teams with diverse life experiences in developing innovative solutions to engineering problems and the disadvantages (opportunity costs in designs not produced) of nondiverse engineering teams. As Dr. Wulf reminded us, engineers are charged with developing elegant solutions that satisfy a variety of constraints, and the more different perspectives that can be brought to bear on a problem, the higher the probability of identifying the optimal solution.

The opening panel of corporate executives presented examples of successful diversity programs developed by different types of employers. Mary Mattis of Catalyst, a nonprofit organization that promotes women in business, provided an introductory overview of corporate diversity management programs. Panel members represented a civil engineering consulting firm (CH2M HILL), an electronics design and manufacturing firm (Motorola), and a utility company (Consolidated Edison).

The participants then reconvened in small groups to discuss the meaning of best practices and how to define success. Their discussions focused on specific programs for improving the recruitment, retention, and advancement of women and minorities, as well as the importance of leadership and commitment from all levels of management. A common theme of the breakout discussions was that finding and keeping good technical employees is critical to the success of a company and diversity programs contribute to a company's ability to maintain a strong workforce.

After lunch a second panel, made up of representatives of minority engineering societies, presented employees' perspectives on corporate management of diversity. Leading spokespersons for the National Society of Black Engineers, the Society of Women Engineers, the Society of Hispanic Professional Engineers, and the American Indian Science and Engineering Society provided a variety of views on the impact of corporate diversity programs and the roles of associations in recruiting, retaining, and advancing employees with diverse backgrounds.

Following the second panel, Willie Pearson, a sociologist and chair of the School of History, Technology, and Society at the Georgia Institute of Technology, presented his findings on the career experiences of African-American chemists over the last 50 years and related them to the current challenges facing minority engineers in the workplace. Civil rights legislation in the 1960s had a significant impact on the career opportunities available to African-American scientists, but Dr. Pearson's research shows that even in the 1990s highly qualified minorities faced many challenges in obtaining positions and advancing in their chosen career paths.

These presentations were followed by a second round of breakout discussions that gave participants a chance to focus on four specific issues: effective responses to lawsuits, affirmative action backlash, the components of effective mentoring programs, and diversity in the global marketplace. Speakers introduced each topic and participated in the discussions. Workshop attendees were encouraged to participate in two of the four discussion groups, and facilitators presented summaries of the discussions to the entire group.

The second day of the workshop began with an address by Nick Donofrio, senior vice president of technology and manufacturing at IBM, on his experiences in managing diversity in the technical workforce. Mr. Donofrio's perspective embraced several points of view. He spoke as chairman of the board of the National Action Council for Minorities in Engineering, as an executive with National Engineers Week, and as an engineer and manager at IBM. He emphasized the business need for more technical professionals in the information economy, the importance of the various groups addressing diversity issues working together, and the need for outreach programs to get young people excited about engineering.

During the third and final breakout discussions, participants identified followup activities to address the issues of the recruitment, retention, and advancement of women and minorities in engineering. Although the focus of the workshop was on employment practices, much of the discussion in the final breakout sessions centered on education. Suggestions included sharing the excitement of engineering with young people, educating precollege teachers about what engineers do, ensuring that students receive a strong foundation in reading and comprehension in the early grades, providing skilled math and science teachers in middle schools and high schools, improving the teaching techniques of engineering faculty, and providing financial support for disadvantaged students.

Jim Padilla, vice president of global manufacturing at Ford Motor Company, closed the workshop with a talk on the business case for diversity. As a member of Ford's Executive Council on Diversity and Worklife and executive sponsor of the Ford Hispanic Network Group, Mr. Padilla has led the company's efforts to increase diversity awareness in Ford's manufacturing plants and has participated in outreach efforts to the Hispanic and African-American communities in Detroit. In his remarks, Mr. Padilla emphasized the importance of diverse viewpoints in the development of innovative products and services for a global marketplace.

THE BUSINESS CASE FOR DIVERSITY

During the workshop discussions, a common theme and four common issues emerged. In general, employers acknowledged that policies and programs that allow all employees to succeed in the workplace contribute to corporate success. In addition to this general principle, four issues surfaced repeatedly in discussions of why corporations care about diversity in the engineering workforce. First and foremost was the need for talented workers and the difficulty of finding enough qualified personnel. Workshop participants agreed that, although many corporations are being forced to reduce their workforces in response to current economic conditions, the long-term demand for engineers and other scientific and technical personnel will continue to increase. The United States cannot continue to rely on immigrants to fill engineering jobs in traditional and high-tech industries. A second related issue was the high cost to companies of employee turnover, including the tangible costs of recruiting and training replacements and the intangible costs associated with maintaining good relationships with clients and suppliers.

The third issue was the perceived competitive advantage of having a diverse workforce, which enables a company to provide better service to increasingly diverse clients and markets, a business imperative for companies that provide both engineering services and engineered products. Finally, engineers with different ethnic, gender, and cultural backgrounds bring a variety of life experiences to the workplace that, if wisely managed, can encourage creative approaches to problem solving and design.

KEY COMPONENTS OF SUCCESSFUL CORPORATE DIVERSITY PROGRAMS

Speakers and participants at the workshop discussed many aspects of successful corporate diversity programs. The most frequently mentioned components are listed below:

- High-level commitment. The CEO, senior management, and board of directors of the organization must demonstrate their commitment to workforce diversity, not only by issuing statements and policies, but also by making appropriate decisions and taking appropriate actions.
- Clear link to business strategies. Management must show that workforce diversity helps the organization meet its business goals and is good for the bottom line.
- Sustained effort. Changes in workplace cultures and behaviors do not happen overnight. Diversity programs must take a long-term approach.
- Training. Managers and employees need training to address workplace diversity issues.
- Employee affinity groups. Workshop participants discussed the pros and cons of affinity groups (e.g., African-American employees, Hispanic employees, women, etc.). Most felt that affinity groups help organizations identify issues and communicate more effectively with stakeholders.
- Outreach to the educational system. To increase the pool of future engineers, corporations must develop partnerships with engineering schools

and precollege educational institutions. Young people need a solid grounding in math and science in their elementary, middle, and high school years to have the option of pursuing an engineering career. Corporations can invest in the future by working with the government and universities to strengthen precollege education.
- Accountability. Individuals responsible for implementing diversity programs must be held accountable for results. In the business world, this means linking results and compensation.
- Benchmarking against other organizations. Successful companies keep track of what the competition is doing in terms of diversity, just as they do in terms of other business goals.
- Communication. Frequent and consistent communication about the goals and programs up, down, and across the organization is important to maintain focus and ensure common understanding.
- Expanded pool for recruiting new employees. Companies must sometimes look beyond traditional sources for new workers to increase diversity.
- Monitoring progress. Metrics for determining success in managing diversity can be difficult to define, and companies must consider more than the numerical mix of demographic groups in the workforce. As a rule, what gets measured in corporations gets done, so defining metrics and tracking progress are critical to keeping management attention focused on the issue.
- Evaluating results and modifying policies when necessary. Like an engineering design project, a diversity program needs a feedback loop to ensure that the desired results are being achieved.

Introduction

CORDELL REED
Chair
NAE Committee on Diversity in the Engineering Workforce

I'd like to give you some background on why we decided to hold this workshop and what we want to achieve. First and foremost, because of the leadership of Bill Wulf, the president of the National Academy of Engineering (NAE), increasing the diversity of the engineering workforce is a primary objective of NAE. An important component of NAE's strategy for increasing diversity in the engineering workforce begins with bringing together stakeholders to share their knowledge, identify information and program needs, and initiate actions to address those needs. NAE held a summit on women in engineering in May 1999, which many of you attended. In September 1999, NAE held a workshop to discuss the business case for diversity. As a result of those two workshops, we found there was a large, enthusiastic group of people anxious to deal with this subject. To follow up on what was learned at those workshops, NAE established the Forum on Diversity in the Engineering Workforce to bring together government, industry, education, and academic stakeholders to review existing information and to define and initiate programs. NAE also appointed the Committee on Diversity in the Engineering Workforce, of which I am the chair. The committee puts on workshops, makes recommendations, conducts studies, and generally serves as a vehicle for taking action. The Forum and the Committee on Diversity in the Engineering Workforce hoped we could make an immediate impact by holding this workshop to share our experiences on the effective management of diversity in the workforce. We are going to discuss the retention, recruitment, and advancement of people with diverse backgrounds in technical careers. The proceedings will be published by NAE in both electronic form and traditional paper form so that a wider audience can benefit from our experiences.

Representatives of companies that have dealt successfully with diversity will describe their experiences and back up their claims with data. We'll take a close look at metrics for determining progress and lessons learned. We will also talk about how the experiences of large companies can be used by smaller companies and how experiences in one type of company can be used by another type of company. We will also address some of the hard issues, such as backlash, lawsuits, and the pros and the cons of mentoring. Finally, we will have an opportunity to get feedback from women and minorities who will present their views about how well companies have been doing in their efforts to increase diversity. We hope the results of this workshop will help companies negotiate the present economic downturn and prepare for the economic recovery.

The Importance of Diversity in Engineering

WM. A. WULF
President
National Academy of Engineering

I want to share some thoughts with you from a talk I gave to the NAE Annual Meeting about two years ago, in which I tried to explain why I believe we should be deeply concerned about diversity in the engineering workforce. I feel very deeply about this issue because I believe diversity in the engineering workforce is an absolute necessity. It's not just that it would be *nice* if we were more diverse; the issue is much more important than that. I believe it is an *absolute necessity*.

Many people talk about the need for diversity as an issue of equity, in terms of fairness, and that is a potent argument. Americans are very sensitive to issues of equity and fairness, so the argument resonates with many people. A second argument for diversity has to do with numbers, the fact that white males are becoming a minority in the population of the United States and that, unless we include more women and underrepresented minorities in the engineering workforce, we are simply not going to have enough engineers to continue to enjoy the lifestyle we have enjoyed for the last century or so. This, too, is a potent argument, but it is not the one I am going to present today.

My argument is essentially that the *quality* of engineering is affected by diversity (or the lack of it). To support that argument, I am going to share with you some very deep beliefs about the nature of engineering, some of which run counter to stereotypes of engineers and engineering. The argument in a nutshell hinges on the notion that engineering is a profoundly creative profession—not the stereotype, I know, but something I believe deeply. The psychological literature tells us that creativity is not something that just happens. It is the result of making unexpected connections between things we already know. Hence, creativity depends on our life experiences. Without diversity, the life experiences we bring to an engineering problem are limited. As a consequence, we may not find the best engineering solution. We may not find the *elegant* engineering solution.

As a consequence of a lack of diversity, we pay an opportunity cost, a cost in designs not thought of, in solutions not produced. Opportunity costs are very real but also very hard to measure. The stereotype of engineering in this country does not include a notion of creativity. According to the stereotype, engineers are dull; they are nerds. Unfortunately, I think that is part of the reason we have not achieved the same level of diversity in our profession as we have in the population. We need to break this negative feedback cycle. When I speak of diversity, I mean the kind of inclusion you probably thought of instantly, that is, appropriate representation of women and underrepresented minorities. But my idea of diversity also includes the notion of *individual diversity*, that is, the breadth of experience of a single individual.

When I made this argument to the NAE members a couple of years ago, I had just seen some numbers about engineering enrollments. Undergraduate enrollment in engineering has been dropping since the mid-1980s, about 20 percent from that peak, and down about 3 percent since 1992. Graduate enrollment has been growing, but largely because of an influx of non-U.S. students. In fact, the U.S. student component of undergraduate enrollment is dropping, in spite of the fact that starting salaries for engineering graduates are 50 to 100 percent higher than those of students graduating with bachelor of arts degrees.

My friends who are economists keep telling me that the disparity in salaries will eventually motivate more students to go into engineering. But that is not what the data show. We need to stand back and ask ourselves why, in a society that is *so* dependent on technology, in fact, in some ways is *addicted* to technological change, and with 50 to 100 percent disparities in salaries, engineering is not an *attractive* discipline. Specifically, we must ask why it isn't attractive to underrepresented minorities and women. Traditionally, engineering was thought of as a way to higher economic status. That was certainly true in my generation, but it seems not to be the case now. We need to stand back and ask ourselves why.

Even more disturbing than the overall numbers are the numbers for underrepresented minorities and women. I told you that overall enrollment has dropped 3 percent since 1992, but minority enrollment has dropped 9 percent! African-American enrollment has dropped 17 percent! The percentage of women has held steady, just a tad under 20 percent of the entering freshman class, but those numbers, bad as they are, don't tell the full story. At the same time the number of engineering students has been going down or holding steady, the number of minorities entering universities has been going up, and the number of women entering universities has been going up. That means engineering is capturing a smaller and smaller *market share* of the total enrollment.

The situation is different elsewhere in the world. There is something uniquely Western (except for France) about these numbers. A few years ago I toured

Taiwanese universities, where 35 percent of the undergraduates are engineering students. Forty-six percent of mainland Chinese undergraduates are studying engineering. Half the people at the ministerial level in Taiwan have degrees in engineering. In this country, only a handful of people in Congress are engineers.

Now let's return to the first part of my argument, creativity. My favorite quick definition of engineering is "design under constraint." We design solutions to human problems, but not just any old solution will do. Our solutions have to satisfy the constraints of cost, weight, size, ergonomics, environmental impact, reliability, safety, manufacturability, repairability, power consumption, heat dissipation—the list goes on and on. Finding an elegant solution that satisfies those constraints is one of the most creative acts I know. Think of the word *elegant* for just a minute. I believe that all great engineering achievements, from the Golden Gate Bridge to Post-It notes, are elegant. They are spare. In Einstein's words, they are "as simple as possible, but no simpler." They are aesthetically pleasing, and they appeal to our humanity. They are humane!

Let me tell you a personal story about creativity and elegance. My father and my uncles were engineers. So, in a sense, I was programmed to become an engineer, too. I never seriously thought of pursuing any other career when I went to college. However, I can tell you the exact moment I became *hooked* on engineering. Between my sophomore and junior years at the University of Illinois, Chicago, I was working for Teletype Corporation as a draftsman. My job was doing inking on vellum, the most awful job in the entire world! If there is any job designed specifically to turn people off of engineering, it is inking on vellum. The team I was attached to was, among other things, designing an automatic telephone dialing device. A little punched plastic card was inserted and little mechanical feelers came out and sensed where the holes were and dialed a telephone number. Occasionally, when the cards went through the reader they jammed.

I was hooked on engineering the moment I looked up from my drafting table at the dialer and saw what the problem was. I suddenly understood, and I understood the elegant solution to the problem. I mean, *really, really* elegant! I made a mock up of the solution with a bit of cardboard and drafting tape, and it worked! My boss then had some metal parts made, for a total cost of pennies. It was really exciting. More senior engineers who had been fiddling with this problem for a long time praised me, and I got a bonus in my paycheck. For years, I thought about the fact that thousands of people around the world were using this dialer with no problem with binding. They may have had other problems, but they didn't have a problem with binding. That was all neat!

But what *hooked* me was the moment I looked up and saw the elegant solution, that moment of creativity. Looking back on my career, I have been fortunate to have had that experience a number of times. I can vividly recall each and every one of them because that is what engineering is all about.

Sam Florman, a member of the NAE, wrote a book called *The Existential*

Pleasures of Engineering (St. Martin's Press, 1976) that makes the same point. Florman talks about the *joy* of creation, about the fact that creativity is what makes engineering an *interesting* profession. He cites a psychological study that had been done a number of years earlier that describes engineers as "intelligent, energetic, unassuming people who seek interesting work." Note that they seek interesting work, not dull, pocket-protector stuff; interesting work, work that in some ways is more closely related to that of our colleagues in the arts than our colleagues in the sciences. As Florman says, "The artist is our cousin, our fellow creator." Bob Frosh, another NAE member and a former administrator of NASA, sent me a quote from the editor of the codices of Leonardo da Vinci. Talking about the impact of editing the codices, he said, "At last people will start believing me. da Vinci was an engineer who occasionally painted pictures when he was broke."

The point is that engineering and art are not on opposite ends of a spectrum. They are, in fact, closely related to each other. Indeed, a defining aspect of human beings is the use of tools to modify the environment. That is what distinguishes us from the great apes. In this sense, engineering is the most humanistic of all activities. Obviously, engineering also has an analytic side, maybe even a dull side that comes from an innate conservatism. Just like medical doctors, the rule is "first, do no harm." Our conservatism and our creativity are always in tension. Indeed, the most original, most creative design is the one about which we are the most skeptical. If you make small, incremental changes in previous designs, you don't meet much resistance. But really creative, far-out designs arouse great concern. That is why, immediately after our most creative moments, we begin looking for flaws. We put on our skeptics' hats and subject our ideas to careful scrutiny, trying to ferret out the possible downsides, all of the ways the designs might fail. In short, instead of celebrating our creations, we try to find their flaws. To meet our responsibilities, that is exactly what we ought to do.

Unfortunately, that is the only side of engineering the public sees. To quote Florman again, "it is especially dismaying to see engineers contributing to their own caricature." I can easily get a laugh out of an audience of engineers by describing them as white-socks, pocket-protector, cubicle folks. I think that caricature is one of the biggest problems keeping young people from pursuing careers in engineering, despite the fact that study after study after study has shown that both women and underrepresented minorities are attracted to professions in which they can contribute directly to the welfare of others. That is why we find more parity in the legal profession and in the life sciences. But, in *fact*, engineers have contributed more to the quality of life than those other professions.

No one seems to think of painters or artists as dull people, but think about how long Michelangelo laid on his back painting the ceiling of the Sistine Chapel, the brute strength it took to plaster that ceiling while lying down—not a very exciting activity. A friend of mine who is an Emmy award-winning director set up a weekend for me with a group of Hollywood film makers to see if we could convince them to produce a show called "L.A. Engineer." It turned out we could not, but one of the things that I learned that weekend was just how dull it is to make movies! The actual shooting time of the movie is very brief; but months and months are then spent in a dark studio editing this film. It is really, really dull. Every profession, whether painting the Sistine Chapel or making "L.A. Law" or being an engineer, has a creative side and a dull side. To increase our diversity we must make young people want to be engineers, and to do that we must change the stereotype.

Now I want to turn to my second theme, diversity. I repeat the simple truth that creativity is bounded by life experiences. The psychological literature is very clear about this. Creativity is simply making unexpected connections between things we already know. If engineers were as dull as they are in the popular stereotype, they wouldn't be good engineers. They wouldn't have the life experiences they need to come up with creative solutions to human problems. Let me repeat. If engineers were really as dull as the stereotype, they wouldn't be good engineers!

As president of the NAE, whose members are among the most creative engineers in the world, I can tell you they are *really* interesting, and that is not a happenstance. Collective diversity, what people usually mean by diversity, is essential to good engineering at a fundamental level. Men, women, people from different ethnic backgrounds, the handicapped—each of them experiences a different world. Each of them has had different life experiences.

I think of these life experiences as the "gene pool" out of which creativity comes, out of which elegant engineering solutions come. The quality of engineering is affected directly by the degree of diversity in the engineering team for that project. It doesn't take a genius to see that, in a world of global commerce, we must design products that are sensitive to many cultural taboos and for very different customers. But the need is deeper than that. The range of possible solutions to an engineering problem will be smaller from a nondiverse design team, and the *elegant* solution to a human problem may not be among them. That limitation can have substantial, but hidden, economic costs, opportunity costs, costs that must be measured in terms of designs not considered.

Opportunity costs are very hard to measure, but they are very real. To illustrate the problem, let me tell you something from my own experience. One of my interests over the years has been computer security, and until fairly recently, I still had two graduate students at the University of Virginia. One of my students came to me with a problem she wanted to solve. She wanted to be able to run an application program and to know that either (1) the application had not been compromised and was, therefore, working correctly or (2) that it had been compromised and should be ignored. But she wanted to run this program on a computer belonging to the "bad guys" who had access to everything. They could pull the plug out of the wall, could examine all of the software including the software my student wanted to run, could make arbitrary modifications to the underlying operating system, could make arbitrary modifications to the hardware, could modify the application my student wrote, and so on. In addition, because the application had to run virtually forever, the bad guy would have all the time in the world to analyze the situation. I looked at that problem, and I said, "No way. You can't do that!" I told her not to waste her time, that it was an impossible problem.

But my student found a solution; not just any old solution but a truly *elegant* solution. I don't know whether it was because she is a woman or because of her Chinese background, but her life experiences enabled her to see a solution I would never have seen. Once she explained it to me, I understood it, of course. In fact, I was able to build a proof that it would work—a nice linear, male, left-brain proof.

Now let me bring the themes of creativity and diversity together. I believe that a central factor in the declining enrollment in engineering, especially the declining enrollment among women and underrepresented minorities, is the stereotypical image of engineers. We know there are also many other problems, of course—the lack of mentors, the lack of family support, the absence of role models. We could put together a long list of problems. But to my mind, they don't explain the declining enrollment. It must be these kids don't want to be engineers! There is something about engineering that is vaguely repugnant to them, and we need to understand what that is. There may be several things, but one of them is certainly the image. What really bothers me is that the image is incorrect! Engineering is *not* dull. Engineering, in fact, is an enormously fun, creative, rewarding profession that has had a profound impact on the quality of human life.

The image of engineers is very different in other places, in France, for example, and China, both Taiwan and mainland China. In fact, in this country from the early nineteenth to the mid-twentieth century, engineers were celebrated as heroes in books, in film, and in poetry. Consider a few of the many quotations about engineers: Walt Whitman, "Singing the great achievements of today, singing the strong light works of engineers," or Robert Louis Stevenson who wrote about the engineering of the transcontinental railroad, "If it be romance, if it be contrast, if it be heroism we require, what was Troy to this?" I could cite dozens of other examples.

The nerdy image of engineers is not ordained. It is not ordained that the contributions of engineers to our society be discounted. It is not ordained that our image remain repulsive to the diverse students we must reach for excellence in engineering. The NAE has initiated a number of programs to address these issues, and this workshop is an essential component of those programs.

To sum up, I believe that diversity is essential to good engineering! In addition to the issue of fairness and equity, in addition to the issue of numbers, there is an issue of *quality*. For good engineering, we *require* a diverse engineering team. But for some reason, engineering has become repugnant to young people. We need to face that fact and try to change it. There is no silver bullet to fix our image. We are going to have to work on it over a long period of time. But if we don't start working on it, we're never going to break out of this destructive, negative feedback cycle. In the meantime, we can do a great deal. The organizations you represent have already taken aggressive, visible actions to address the problem. I believe we can make a start by sharing your experiences with each other and with us.

Best Practices in Managing Diversity
A Panel Discussion

Introduction

MARY C. MATTIS
Senior Research Fellow
Catalyst

My comments are intended to provide a brief overview of corporate best practices in managing diversity and establish a framework for exploring diversity initiatives represented by our panel today. Catalyst, a nonprofit organization that works with corporations and professional firms to retain and advance women, has been in business for almost 40 years. In the beginning, we worked with individual women, but we now work almost exclusively with corporations or professional firms to help them retain and advance women. We think women have done all of the right things—they have the right educational credentials, and they are getting the right experience. Now we need to fix companies.

One of our activities is the Catalyst Award, an annual process during which we ask companies to nominate a specific, cutting-edge, replicable initiative that can be proven to have been successful in retaining, developing, and advancing women. We spend about a year evaluating the nominations; we visit the companies and talk to many people, including the CEO, women, human resources professionals, and so on. At the end of this very demanding process, we select two or three companies that we hold up to the public as companies whose initiatives are working. Most diversity initiatives are not targeted only on women. In most cases, these initiatives not only help the company retain and advance talented women, but also retain and advance members of other minority groups. All of the best practices we have reviewed recently are multifaceted because there is a growing recognition in the corporate community that diversity is multifaceted. There is no single solution, no one-size-fits-all approach that addresses the needs and interests in our diverse workforce.

I had nothing to do with the selection of today's panelists, but two of the companies represented on this panel have won the Catalyst Award—Con Edison, for a very interesting program that ensures that new people hired in engineering, as well as in business, rotate from headquarters to field locations to ensure that they receive both line and staff experience, and Motorola, for a very interesting succession planning process that ensures that women and people of color are represented in succession planning.

Companies use a variety of business-based arguments as a basis for developing, implementing, and driving diversity initiatives through their organizations. One example of the basic business case for diversity is demographics. The talent pool, both today and in the future, will become increasingly diverse in terms of gender, race, ethnicity, and global representation. The real competitive advantage for a company out to get the best and brightest talent, especially technical talent, which includes a very limited number of people from diverse domestic groups, is to be regarded by prospective employees as the employer of choice.

Another business case for diversity that many employers now recognize is the cost of not acting, the cost of turnover if the environment is not welcoming for people from diverse backgrounds. Some of the costs are obvious, and some are not. They include replacement costs, opportunity costs, which Dr. Wulf talked about, the impact on work units, customers, and clients, and the discontinuity of service. Clients and customers dislike having to work continually with new people, a company with a revolving door. If you can't retain talented people, the word gets out, and eventually you can't recruit them to your organization. If a company has to respond to litigation, we all know there are significant costs associated with that, not just in terms of real dollars, but also in terms of image, opportunity costs, and the impact on recruitment. As Dr. Wulf so eloquently explained, one of the strongest business cases for diversity is that diversity equals innovation—in terms of new products, new services, new markets—and, of course, the bottom line—increased profitability, increased market share, more elegant solutions.

Based on our research on corporate best practices for retaining and advancing women, and also other minorities, we have identified the following characteristics of successful corporate initiatives. First and foremost is top-level commitment. Diversity initiatives cannot be driven through an organization by human resources people alone. There must be a commitment from the CEO and senior line managers throughout the organization.

Second, the rationale for diversity initiatives must be linked to business strategies. People have to understand the company-specific business case for them to support the initiative all down the line. Communication of the business case for diversity, as a CEO said to me once, takes a million messages. Companies must use every possible vehicle to remind people why diversity is important and then follow up with training to teach people how to behave appropriately.

Third, the company must use internal and external benchmarking to develop metrics—for early wins as well as long-term goals. Ongoing monitoring must be a long-term process. One of the metrics must be accountability, so that people understand they are just as accountable for diversity results as they are for other business results. Finally there must be ongoing evaluation to determine what works and what should be changed.

We have also identified some trends in corporate diversity initiatives. First, companies are increasingly benchmarking what other companies are doing with diversity initiatives just as they have traditionally benchmarked other kinds of "hard" business results. Companies are trying to learn from the experiences of others, as opposed to reinventing the wheel. Second, companies are adopting formal, rather than ad hoc, approaches. In other words, they formally inform people of the policies and programs to make sure the word gets out that the company values diversity, the company does the kinds of things that attract and retain a diverse workforce, the company is a multicultural company as opposed to a one-size-fits-all company. Third, companies are pursuing diverse workers for business reasons, rather than for regulatory compliance. Companies communicate to their employees, and to their potential employees, that they are acting in response to a business imperative and not because someone out there is counting the numbers.

Fourth, companies are working through partnerships, rather than alone. Companies are beginning to work together, even with competitors, to accomplish things they couldn't afford to do alone. A good example is IBM's American Business Collaboration for Quality Dependent Care, which initially collected a pool of $25 million from a number of companies to improve dependent care facilities in the communities where they operate. Partnerships focused on diversity are also increasing. Finally, companies are adopting a truly *diverse* model, rather than an *assimilation* model. In essense, they tell new employees that they want them to bring their whole selves to work. They are looking for whole individuals, not cookie-cutter copies.

Diversity Practices at CH2M HILL

DANIEL E. ARVIZU[1]
Group Vice President
CH2M HILL

Good morning. I'm Dan Arvizu, and I'm pleased to be here to talk about a subject that's very near and dear to my heart, a topic I spend a lot of time thinking about. I hope I won't disappoint you in not having a plethora of statistics to go with this talk, but I think some of the themes that have already been set up in the opening introduction and in Dr. Wulf's talk speak to the nature of the diversity initiatives in corporate America.

Let me start by giving you a very quick profile of our company to help you understand where I'm coming from. CH2M HILL is a global project delivery company with 11,000 employees. We have 120 offices—actually, 200 area offices, but 120 permanent offices—on seven continents. We provide engineering consulting and other integrated services to a variety of clients, and we generate about $2 billion in revenue. What is interesting to me about this company as it relates to diversity is that we've roughly doubled in size in the last five years and plan to double again in the next four to five years. So this is a very robust, growing company and one that I believe has a lot of interesting features.

One of the things that attracted me to the company—I've been with the firm for about three years after a long career in the national laboratory system—is that it's employee owned. That's an important facet of this discussion because employee ownership means that I'm a shareholder in my company. As a result, I hold my management accountable for the decisions they make. And those decisions clearly affect everything about the company, including, but not limited to, the bottom line. In our business practices, we're always looking for the business case and asking why this is important and how it reflects on the bottom line.

---

[1]Dr. Arvizu's remarks are used with permission.

An employee-owned company has a strong advantage in that it isn't publicly traded. This gives us a bit longer time horizon in which to operate. If we're focused on growing and doubling again in the next four or five years, what we're really focusing on is how to attract the best and the brightest people. To meet our financial targets, we have to add a person a day. That's 300 or 400 people per year, a lot of employees to try to hire, especially in a profession with a daunting war for talent.

Our history of growth and the fact that we're still in a growth mode are relevant to the topic at hand. The diversity initiative at CH2M HILL is designed to help us understand and attract the best talent we can find and retain the talent we have. Retention is a big issue. It costs our company between $30,000 and $60,000 to replace an employee who leaves. So it's much cheaper for us to retain an employee that we've trained and started on a career path with the firm than it is to go out and replace employees when they leave.

This company, because it's made up of consultants with diverse backgrounds who are located around the country and across the globe, has a lot of working relationships in the communities where we live and work. Because our clients reflect the demographics of the nation and of the world, our employee profile has to reflect that as well. We never lose sight of what we're trying to do—to add value both to the shareholders of our company and to the clients we serve.

So what do we do? Throughout its 50-year history, CH2M HILL has traditionally focused on hiring at the entry level, then developing and promoting people from within. As part of this philosophy, we have implemented work-study programs to identify the cream of the crop of future employees in the pipeline of universities and the local communities in which we work.

We consider career development very important, so we spend a lot of time on mentoring and on moving professional staff laterally, giving them the opportunity to gain experience throughout the company. Formal career goals, succession planning, and evaluation of employee interest and track records are part of this process. We furnish training in the skills required for the company to benefit from diversity and also to foster understanding of why diversity is so important. Along with internal training, the company supports continuing education for individuals who want to add a particular component to their skill mix.

We established a succession management process at various levels in the company. These "succession pools" must include a diversity component. We measure and link the performance of people in meeting diversity targets to their compensation. We have a unique internal Web system where every employee can go in and learn about their own personal financial metrics, such as client ratio and overhead cost structures in particular groups, but also the diversity statistics that go along with all of those things.

We also do benchmarking to assess how we're doing in relation to our industry and to the rest of the world. For example, we identified six companies we think are a lot like us and looked at what they do. We reviewed their best practices and tried to emulate the things that work and learn from the things that don't. Some of the firms we benchmarked against include PG&E (San Francisco), American Management Associates (New York), Northwestern Mutual Life (Milwaukee), CompUSA (Dallas), and USA Group (Indianapolis). Because these companies are progressive in diversity, they're important to us. Benchmarking to learn from what others are doing takes sustained effort. To truly integrate diversity, you can't just roll out a couple of initiatives and expect to be done. Diversity has to be part of the overall effort that makes your business tick. You have to do what's important in your particular organization.

In addition to looking at other companies, we do lots of voluntary employee surveys. On our internal Web site, people can log in and complete these surveys, and although the surveys aren't mandatory, employees do respond. We also hold focus groups, which provide opportunities for people to offer their individual comments. We use similar techniques to assess staff attitudes. We also try to understand why people leave by talking to objective third parties. We continue to provide mechanisms by which people can give us feedback.

We take a fairly aggressive, essentially analytical approach to understanding the internal statistics. These data become part of our spreadsheets, if you will. We have a lot of engineers, of course, who are very analytical. Everybody is always asking for the data. Although we do make the data available, we try not to get hung up on the numbers because the issue of diversity has many different elements. However, data management is something we do quite actively.

And finally, we have a board of directors committee on diversity in the workforce, on which I serve. Diversity is an important piece of our corporate culture. One of the main ingredients in best practices is commitment at the top. Our CEO, Ralph Peterson, is passionate about this topic, and he supports it personally with his time and involvement, as well as through corporate sponsorship.

I can't overemphasize the importance of the relationships the firm maintains with various other entities that focus on educational and skill training opportunities. Historically black colleges and universities (HBCUs), of course, are a very important mechanism for identifying top talent. Our CEO maintains a strong relationship with HBCUs. The Hispanic Engineer National Achievement Award Conference Corporation (HENAACC) is another organization that our firm feels strongly about; we support HENAACC both with sponsorship and on an individual level. The company directly supports employee associations and diversity councils, and we've established programmatic ties with placement groups, academic programs, and trade schools to broaden our hiring pool. We use various other mechanisms to help us connect to the local community and to demonstrate our strong commitment to the underrepresented groups in the engineering field.

How do we track our progress? Metrics are an important tool. We are a very diverse company with a lot of small offices around the world. Our office in Boise, Idaho, for example, serves clients who are very different from our clients in Los Angeles or Atlanta. Our clients expect a certain mirroring of their demographics and their particular communities. So we use metrics on a local basis, as well as on a broader basis. Our workforce comprises roughly 22 percent women and 8 to 9 percent underrepresented minority ethnic groups. Although that represents the national average in terms of availability for the kinds of professionals we seek, it doesn't necessarily represent the demographics of a local community. That's why we try to focus in on particular areas.

In our industry, retention statistics indicate roughly 20 percent turnover. At CH2M HILL we do a bit better than that. However, turnover is very expensive for companies and has a strong impact on the bottom line. When economic times get tough, as they are now, turnover tends to fall. But, historically, turnover even in the low teens in terms of percentage, remains a very important area of corporate focus. Although we do better than our industry average, we need to do even better.

From my career in the national laboratory system, with brief exposures to AT&T and Lockheed Martin, I've learned that corporate practices vary widely, and every business is different. Although there is no silver bullet, a number of different things work in a particular set of circumstances. Here are some of the critical success factors with which we challenge our people. We have a clear vision that says we reflect the demographics of the clients we serve and that we bring the diversity and the broad suite of engineering talent and solutions to clients with very interesting and important problems.

We articulate our business case. One of the founding fathers of CH2M HILL, Jim Howland, made it clear from the beginning in his "Little Yellow Book," which you can find on our Web site at CH2M.com. Howland described what the founders envisioned for this company. He suggested that the firm should provide an opportunity for engineers to get together to do interesting work, enjoy the people they work with, make a little money, and have some fun. That philosophy reflects the ethos of this firm, and that's how we try to attract and keep people engaged in the business.

Another critical success factor that may be missed sometimes is that we specifically define what diversity is and carry on a continuing dialogue about it. We create a common, inclusive language that helps us engage in that discussion.

Then we make diversity everybody's job. Although strong leadership support is clearly important, everybody must "own" diversity because it contributes to shareholder value. There may be individuals in the firm who are very entrenched in their ways, and you have to bring them along with the rest of the group. That requires change in the *culture* of the organization.

In the long term, incorporating diversity requires courage. It requires service leadership. It isn't just a box on some form somewhere that you can check. It requires sustained effort. You must continually nurture and mentor *everyone* around you. And that is how we look at it. That's how we challenge our own folks, and that's our commitment. Our diversity strategies are linked to our core business strategies of developing leaders, planning for successors, strengthening the role of group leaders, developing local community relationships, and growing the company both in the United States and around the globe. Diversity and the issues that relate to it are tightly woven into the very fabric of our firm.

Diversity Practices at Motorola

IWONA TURLIK
Corporate Vice President and Director
Motorola Advanced Technology Center

I would like to share with you my perspective as an engineer and manager at Motorola where we try to go beyond the numbers that represent the traditional definitions of diversity. Our definition takes into account a wide range of human attributes, such as thinking styles, age, experience, education, geographic origin, economic status, language, religion, lifestyles, and sexual orientation, as well as gender, ethnicity, and disability. As the manager of facilities in Brazil, Germany, China, and the United States, I know that perspectives on diversity are very closely related to culture. Today, I will focus on diversity from the perspective of a manager working in the United States for a U.S. corporation.

Our goal at Motorola is to create a globally diverse business environment and to be recognized by our customers, shareholders, employees, and communities as the premier company for which to work, from which to buy products, and in which to invest. Having a diverse organization results in a stronger business. Diversity is not about employing more women; it is about being a better company that can find better solutions. At Motorola we appreciate different ways of looking at an issue. Our mission is very simple: to ensure the long-term success of the company by empowering "Motorolans" with diverse backgrounds, styles, cultures, and abilities to turn global diversity into a competitive advantage. To be a successful company (after all, what really counts is what your stakeholders think about your business), you need experienced employees with diverse backgrounds.

What is Motorola's alignment and talent strategy? If you look at the list in Table 1, you'll notice that it is not much different from what other companies do. We focus on benefits, awards, and retention. With a diverse workforce, managing employee benefits is not a simple issue because you must offer different packages

TABLE 1 Motorola Alignment and Talent Strategy Framework

Global marketplace
Advancement of women and minorities (integration at all levels)
Diversity of management team
Cultural awareness and acceptance
Integration of individuals with disabilities
Work/life integration
Benchmarking
Excellence as a measure to evaluate global diversity
Equity in expansions or contractions of the workforce
Understanding of local markets and quick action in attracting talent
  (localization of market demand)
Mentoring/networking
Employee wellness program
Measurement and management accountability
Talent management
Talent pipeline initiative
Supplier diversity programs
Minority conferences/recruitment
Motorola business councils
Work policies such as flexible schedules, telecommuting
Benefits
Awards
Retention
Training
Communication strategy
Community involvement
Motorola Foundation grants

---

to different employee populations. Awards are a way of encouraging people to stay, but if one focuses only on awards, retention can be a very expensive proposition. Today it is very important to ensure equity when business conditions require the expansion or contraction of the workforce. If you think about the challenges technology businesses, such as Motorola, are currently undergoing, it becomes apparent that the makeup of the engineering workforce is a critical factor.

Why do we focus on women and minorities? For Motorola's engineering community, the answer is very simple. In the next few years, we will need a larger resource pool. Women and minorities are part of the pool from which we will be drawing for the next 10 years to expand our engineering capabilities as a company. We must do the right things in this area to continue to be competitive. A study of Fortune 500 companies by Covenant Investment Management shows that the average annualized return on investment for companies committed to promoting minority and women workers was 18.3 percent over a five-year period, compared with only 7.9 percent for companies with "glass ceilings." That is the bottom line of the business case.

TABLE 2 Greatest Perceived Obstacles to Career Advancement

	Under 30	Age 30 to 39	Age 40 to 49	Age 50 and over
Women				
1	Discrimination	Balancing work/home	Economy	Discrimination
2	Unappreciated	Economy	Discrimination	Poor management
3	Economy/poor management	Discrimination	Balancing work/home	Economy
Men				
1	Economy	Economy	Economy	Economy
2	Poor management	Poor management	Poor management	Poor management
3	Personal skills	Personal direction	Personal drive	Discrimination

SOURCE: National Academy of Engineering. 1999. The Summit on Women in Engineering Resource Book. Washington D.C.: National Academy of Engineering.

The Bureau of Labor Statistics predicts that by 2006 the proportion of white males in the workforce will decline to 44 percent, while the proportion of white females will increase to almost 40 percent. The proportion of African-American females will increase to 6.2 percent, which is 35 percent more than it was in 1976. Latino representation in the workforce will increase to 4.8 percent, which is 78 percent greater than it was in 1976. If we look at population growth, we can see that the numbers vary by ethnic background. This leads us to ask what we can do so that Motorola can grow with these growing populations.

Table 2 shows the different perceptions of females and males about work. Until you understand the differences, it will be difficult to take action to improve the situation. Women in every age category perceive discrimination as a career obstacle. Men over 50 also perceive discrimination as an obstacle. Regardless of whether or not these perceptions are true, they affect the performance of employees in your company.

What do we do at Motorola to improve retention and advancement? First, we try to include diversity in all of our strategic business plans. Catalyst has recognized Motorola for doing a very, very good job in evaluating and promoting people to senior management positions. However, before individuals can be promoted, they have to be prepared to succeed. Training is a very important aspect of an employee's career path. To help us identify and sponsor candidates for development, Motorola established a special human resources task force to develop systems for selecting and promoting qualified individuals, especially minorities. We have found that there is a fine line between giving individuals an opportunity and promoting them too soon, which can have serious negative consequences. Promotions should take place when employees are ready. Promotions must be fair, for both majority and minority employees. More time should be spent preparing minorities for senior positions because there are fewer of them and because they sometimes require different experiences. The network may also be slightly different. In short, we must be careful that we do not promote too soon and that we promote fairly.

What would we like to accomplish through our diversity efforts? If we are successful, the most qualified women and minority engineers will stay with Motorola. Women and minority engineers will rise without barriers through corporate and organizational ranks, based on their personal talents, interests, and ethics. Overall, the number of engineers will remain at a healthy level and will reflect a vibrant, multicultural workforce, and Motorola will be recognized as a leader in providing diversity in the engineering workforce.

What have we done at Motorola? I would like to share with you one of the insights that occurred to me during the meetings of the NAE Forum on Diversity. Being an engineer and knowing the cycle time for an engineering design, I am used to things changing very rapidly. In contrast, cultural change is a slow process. Trying to change the culture of an organization and making a lasting impact means you really have to view things from a different perspective. You have to focus on something and make a start.

We did a focus group study at one of the Motorola labs I manage in the United States. The lab has 75 employees. Thirty-six employees were randomly assigned to be part of five focus groups on diversity-related topics. The first question was: What factors contribute to your overall job satisfaction in the Motorola Advanced Technology Center? The participants had several answers, such as "ideas are always welcome" and "there is a lot of freedom and openness for engineers' creativity." They also mentioned respect for peers and coworkers, the cultural mix, including women, welcoming and friendly people and interaction. Clearly, diversity was important even though no one specifically mentioned "diversity." Being open and including people from different cultures is more important than having specific numbers of women and minorities in your organization.

The second question was: Do we recognize people fairly well? I was very pleased with the answer from my team on this point. My senior management is very sensitive to recognizing people when they make an impact. I think that this is the most important kind of recognition. If you recognize contributions at the right time and in the right place, everybody benefits, and the organization becomes much better.

We also spend a lot of time on personal career growth in the organization. This is a simple business case. It can be compared to an engineering design. If you spend a lot of time up front before you move to the final design, your results will be much, much stronger. I believe in sitting down with my employees and talking about what they would like to have and where they want their careers to go. We all know that career plans usually can't be realized exactly as planned,

TABLE 3 Motorola Diversity Awards

---

100 Best Companies for Working Mothers (1992–2000), *Working Mother*

The Best Companies to Work for and Why (1992–2000), National Society of Black Engineers

Executive Director's Award for Community Partnership (2000), National Society of Black Engineers Bridge Legacy Program

Hispanic Corporate 100 List (2000), *Hispanic Magazine*

Pride in Excellence Employer of the Year Award (2000), Project Equality, Inc.

Secretary of Labor's Opportunity 2000 Award, U.S. Department of Labor

Top 100 Employers of the Class of 2000, *The Black Collegian*

Top 50 Companies for Diversity (2001), DiversityInc.com

Top Gay-Friendly Public Companies in Corporate America for 2001, The Gay Financial Network

America's Top 50 Corporations (2001), Div2000.com

Corporation of the Year (2001), National Society of Hispanic MBAs

2001 "Outie" Award for significant achievement, Out and Equal Workplace Advocates

---

NOTE: For more information on Motorola's awards related to diversity, see <http://www.motorolacareers.com/ufsd2/diversity/awards.cfm>.

but good dialogue is a good way to keep the organization strong and growing. I really liked the last response, which was that there is something very special about our organization. I think the most important thing is that we are really inclusive and open; at the same time, we complete our milestones and meet the goals the corporation sets for us.

We asked the focus groups if they feel that employees, regardless of their gender, ethnicity, or sexual orientation, are treated fairly by peers, supervisors, and managers. One response was that we have the most diverse organization they have seen. Another response was that we have a good mix of people. We don't favor any special group; we favor very good performance. We are not judgmental, and we have an open-door policy. The lab works well because of the diversity and exposure to different cultures. People are open-minded and treat each other with respect. These are simple, small steps an organization can take to be more successful. There is no magic behind it. Responsiveness and strong leadership involve more than simply measuring the minority numbers.

We also asked how they felt about diversity business councils. The response was that they contribute to networking and awareness, but some employees were not sure what the councils do. Employees also wanted to know why only certain groups had business councils. If an employee is already comfortable and functioning well in a diverse environment, he or she might not feel the need to associate with a special group, might not want to be singled out in that way. Other employees thought that diversity councils were mostly social functions but that the opportunity to hear others' perceptions was beneficial. It works both ways. If we didn't have the councils, we wouldn't have the networking. If you look at the end result, it might seem that a small organization that is very diverse might not need the councils. I would say the opposite—in most organizations, business councils play a very important role, and they should be supported.

Finally, we asked our employees for suggestions for improving our diversity programs. Some said we didn't need to do anything more because we are already diverse. In my opinion, however, if you are not moving forward, you are losing ground. The focus group's response indicated that they appreciated our very good, diverse organization but that not all groups in Motorola are as diverse as we are. Thus, there is still work to do. People might think there is not much to do, but I agree with one of the employees who said that we should market our diverse group to other Motorola organizations and to our management. In this way, we can promote the benefits of a diverse environment.

When it comes to diversity and related issues, Motorola is one of the very best companies for which to work, but we are not perfect. Even though we have won many major awards for diversity (Table 3), we still have plenty to do. Considering the number of women and minorities available in the workforce and the number of technologists Motorola will need in the next 10 years, we face a significant challenge.

Diversity Practices at Consolidated Edison

RICHARD P. COWIE
Vice President of Human Resources
Consolidated Edison Company of New York

I am Rich Cowie, the vice president of human resources at Con Edison of New York. As a non-engineer, I am a member of a minority group in the leadership of Con Edison, where almost everyone else at the top of the company is an engineer. Fortunately, our chairman wants a diverse team, so every once in awhile he picks a non-engineer to join the group.

We are very fortunate to have been recognized for our efforts in recruiting and retaining women and minorities in engineering careers. We were the first utility ever to receive Catalyst's national award for advancing the interests of women in business. We have also received recognition from Catalyst for the representation of women in our executive levels and on our board of directors. Our efforts have been recognized by many other organizations as well. For the fourth consecutive year, *Fortune* magazine has listed us as one of the 50 best companies for minorities. We have also been recognized by *Working Woman* magazine as one of the top 25 companies for executive women. *Equal Employment Opportunity Publication* named Con Ed one of the top 50 companies for minority engineers.

Con Edison of New York serves the five boroughs of New York City and the lower portion of Westchester County. We operate in one of the most diverse areas on earth, and we need the best and brightest in engineering talent to address today's problems and tomorrow's challenges. Success in this area is critical for us.

The tragic events in New York City on September 11 resulted in unprecedented challenges for our company and our people. To restore electric, gas, and steam service to lower Manhattan required laying more than 36 miles of temporary cables on the streets of lower Manhattan, repairing and restoring five underground transmission lines, isolating five miles of gas mains and retiring more than one mile of gas mains in the World Trade Center area, installing 16 new gas valves to improve sectionalization and restore service, constructing a gas regulator station in less than 36 hours, and restoring more than 15 miles of steam mains. The estimated cost for the emergency response, temporary restoration, and permanent repairs is $400 million so far, and the final cost may be even higher. The engineering and leadership challenges have been extraordinary, but our people have risen to the occasion.

Our panel has been asked to talk about what works. So, let me describe three things that have worked for us. First, commitment from the top is critical. I believe the first requirement is a firm commitment to drive the effort from the highest levels of the organization, and we have that. Our chairman has long been active in the National Action Council for Minorities in Engineering (NACME) and has served on its board of directors. He has promoted or placed women and minorities in key organizations in Con Edison, and he meets with the New York chapter of the American Association of Blacks in Energy and has been a speaker at its national conferences. A member of our board of directors, a long time proponent of the advancement of minorities in engineering, has served as president and CEO of NACME and is currently the president of the Cooper Union for the Advancement of Science and Art.

As Dr. Wulf said in his talk, we need to increase the pool of candidates. The second thing that is working for us long term is starting as early as possible in the educational system to get women and minorities excited about engineering and technology. For many years at Con Edison, we have been a key sponsor of and participant in programs at the elementary school level, such as "Say Yes to Family Math and Science," in predominantly minority city schools. We also support programs in high schools and colleges. Company members of our internship program and the American Association of Blacks in Energy and our Black Executive Exchange Program are frequent guest lecturers on careers in engineering and technology. We recruit at career fairs at some 30 targeted colleges, and we have a summer employment program heavily geared to women and minorities.

Finally, our GOLD Program, which used to be called our Management Intern Program, is one of the longest running and most successful programs for recruiting and retaining high-potential women and minority engineers. This is the program that was recognized by the Catalyst Award. It used to be a college intern program, but we now call it the GOLD Program (growth opportunities through leadership development). That name was suggested by people who had been through the program who were asked what most enticed them. They didn't like the connotation of interns or continuing students. So, just last year we renamed the program.

In the GOLD program, through a series of practical, rotational job assignments, mentoring, and senior management guidance, GOLD associates tackle challenging supervisory and project-based jobs that provide them with highly valued work experience as they prepare to move into senior management positions. Over the past five years, 30 percent of our GOLD associates have been women, and 55 percent were minorities.

I think the best measure of what works and how well it works is the people in key positions at Con Edison. Four of the 13 members on the board of directors are women, as are our executive vice president and chief financial officer, our senior vice president for gas operations, and four of our 25 vice presidents.

Among minorities, representation includes three out of 13 members of our board of directors, a senior vice president, two of our 25 vice presidents, and our corporate secretary. Obviously, we have made some progress in bringing women and minorities to the top positions in the company, but our real success lies just below that level. Graduates of our GOLD Program are now in critical jobs at the highest level of our organization, prepared to assume key officer positions in the future. Some of the top positions now occupied by GOLD associates include the vice president for maintenance and construction, the chief distribution engineer, the general manager of substation operations, three plant managers, two electric general managers, the general manager of gas operations, and the director of facilities management.

We have gotten this far by continuous self-criticism. We continually assess how we are doing and look outside our company to identify best practices. In that spirit, I look forward to hearing the thoughts and experiences of our audience.

Question and Answer Session

Lisa Gutierrez (Los Alamos National Laboratory): This question is for Richard Cowie with Con Edison. I understand that the GOLD Program is a development and succession-planning tool and that you have filtered these employees through different experiences. What are the selection criteria for getting into the GOLD Program?

Richard Cowie (Con Edison): We select from our new college hires after trying to identify their potential for future leadership positions—good engineers, to start with, good people skills, the ability to work well in teams—those kinds of characteristics.

Lisa Gutierrez: Is the program available to current employees, for example, who want to go to college? Say they are technicians, and they eventually want to move up in the ranks.

Richard Cowie: We have taken very few from our existing workforce into the program. But we have taken some, and some have done extremely well.

Gary Downey (Virginia Tech): I would like to ask the panel to consider a potential problem with proprietary knowledge in the sharing of diversity practices. In the 1980s when the Cold War was winding down and the competitiveness with Japan, perceived as America's new foe, was rising, there was a movement in the United States to build joint ventures among government, industry, and universities. These eventually fell apart, and it seems to me that the key reason was because of the issue of proprietary knowledge. The government was able to adjust, and universities were able to adjust, but industries were not able to adjust to the challenge of sharing knowledge that had potential design implications.

In the case of diversity, as we move from justification on the grounds of equity and access to justification on the grounds of business and competitive advantage through diversity practices, we could run into the same problem. The NAE's Diversity Forum is considering building a diversity resources Web site, and a number of you have reported on programs in your organizations. To what extent would you be willing to make the details of those projects, the specifics of the surveys, the specifics of the metrics you use to measure what is going on inside your organization, available in a public way? At what point do you run into the problem of proprietary knowledge?

Dan Arvizu (CH2M HILL): That is actually an issue of great importance to my company, which is in a very competitive industry. In the consulting arena, a number of competitors are going after the same talent pool, and we worry about precisely the issue you are talking about. Clearly, this is one of those times when you simply have to put parochial interests aside, at least for the moment, while you attack a much bigger and greater issue. We have tried to address the issue of sharing information directly. We frequently contribute data in the *Engineering News Record* and a variety of other publications as part of aggregate information by industry sectors. No single company's name is highlighted as doing either extremely badly or extremely well. We have tried to get together with a half dozen or more of our staunch competitors and ask them if they would be willing to share information in a blind pool so the industry statistics can be advertised and examined by anyone interested in learning more about the process. Clearly, we are not going to share our internal statistics, but there is a way to get to that level of detail through an objective third party. That way, you remain essentially one step removed.

Mary Mattis (Catalyst): Catalyst has conducted some blind studies for industries, for example when a company has come to us and said they wanted to know how they were doing compared to a competitor, but they didn't want to identify themselves. We got 10 or 12 companies to participate on the basis of confidentiality, anonymity, but it was hard work because there are a lot of legal considerations. Data sometimes has to be destroyed immediately after it is collected. There are a lot of challenges to doing this, but it is definitely worth doing.

Dundee Holt (NACME): Is there anything the panelists can share with us about the allocation of resources? I'm interested in how much companies allocate for precollege, higher education, and the current workforce. Mr. Cowie stated that Con Edison allocates some resources for precollege education. Only about 6 percent of kids want to go into engineering, and we are concerned about building that number up. How do you decide?

Iwona Turlik (Motorola Advanced Technology Center): You decide based on your priorities. Motorola spends a lot of energy on K through 12 programs like Junior Achievement. We also pay a lot of attention to retention. I think the weakest part of our program is with encouraging students at the college level. We spend a lot of time on kids going into college but not as much on retaining them in college.

Richard Cowie: Our chairman, Gene McGrath, really focuses on early education. He is a visiting principle at Washington Irving High School, right down the street from our company headquarters. He has a math tutoring program at that high school that he encourages all of us, very specifically, to get involved in. I don't think we put up a lot of dollars, but we spend a lot of time on those programs.

Karl Pister (University of California): We have been focusing on the issue of changing or creating a positive corporate culture, and that problem, in my view, is embedded in a much larger problem of changing societal culture. If one were looking for the creative solution or the elegant solution, if the energy of corporate America were applied to the problem of changing national culture, would it lessen the corporate cultural problem?

I think this is a reasonable question to ask at this point. I am basing it on the culture of my own state of California, where approximately half of the children of color are educated by teachers who are unprepared to teach in the classroom and are teaching in crummy schools without necessary supplies. That is a much deeper issue, of course, but I think it should be mentioned.

Iwona Turlik: I would have to agree with you. During my presentation, I said that diversity and culture go hand in hand. But can we resolve all of the problems? No. I think the best approach is to take small steps, and in the end, we will accomplish great goals. Focusing on small cultural changes will be much more beneficial in the long term than what we do in the corporate world. But in the corporate world, we have to give individuals hope and a desire to go into engineering. In the past, people associated an engineer with somebody like Edison, and people really wanted to emulate that. Today engineering is considered less glamorous than it was in the past. Highlighting the benefits and glories of the engineering profession is as important as focusing on the cultural change in the beginning.

Dan Arvizu: That is a very important issue that has important implications for all of corporate America. Most large corporations, and many small and medium-sized ones as well, are very anxious to plug into initiatives and programs that address the systemic issues. Corporations will not take on these issues as their own particular vendetta, although there are some exceptions. In fact, I see Ray Mellado, the founder and chairman of the Hispanic Engineer

National Achievement Award Conference (HENAAC) in the audience. HENAAC has a program, in partnership with IBM and some other large corporations, to do specifically what you are talking about, not only to prepare teachers, but also to prepare the families of small children that don't have access to computers and other things like that.

There are programs you can plug into as a corporation that fit those interests and objectives precisely. Many corporations like ours have foundations that can be geared toward programs for younger children. Most of our business processes are aimed at the upper level and retention, but not as much at K through 12 education and family and societal issues, even though those are really important. You have to do it all. You can't ignore any aspect of the problem.

Tyra Simpkins (*U.S. Black Engineer and Information Technology* and *U.S. Hispanic Engineer*): There has been a lot of talk about the image of engineers and how to go about changing that image, as well as about K through 12 programs. Dr. Turlik, I would like more information on the programs your company works on with elementary and middle school students. How can interested schools participate in programs like that? Does your company participate in outreach programs? If you are going to start to change the image, it's important to do that by way of exposure. The more exposure, the better children and other minority groups can be made aware of their choices. Those choices will set the tone for the future.

Iwona Turlik: Motorola is involved in numerous programs. I don't remember all of them, but I think there are 20 pages of programs we participate in since we are a global company. The most important one I recognize is Junior Achievement.

Motorola has received many awards for making contributions to society. One has to look at those awards in two ways. First, the company is making a contribution and doing a very good job in those areas. Second, the company is marketing those programs because they make a difference. It is important to let the public know how you are involved and what makes that program successful.

Richard Cowie: I think programs like tutoring programs, one on one, with women and minority engineers going into the schools and working directly with students, create positive role models that can directly influence students to move in that direction for their education and their careers.

Jim West (Bell Labs/Lucent Technologies): My career was very much affected by mentors. No one really concentrated on mentoring in the presentations. I would like any member of the panel to discuss experiences with mentoring, especially for minorities and women engineers.

Iwona Turlik: We have had a formal mentoring program in one of our businesses for the last five years, and the whole corporation is now trying to emulate that. But mentoring is a very special relationship, and there must be a careful selection of mentors. A mentor may not always be somebody in your corporation. It could very well be someone outside your corporation, even your grandmother. Mentoring is not a simple solution for a corporation. Do you need a formal mentoring program to improve the development of employees? Obviously, if a program is working well, it will benefit the organization. Rather than institutionalizing a mentoring program, an alternative approach could be a networking program.

Richard Cowie: Mentoring is another thing that can be influenced from the top. Our chairman selects people to be his assistants. He includes women and minority engineers in our GOLD Program in the pool of people he looks to. He encourages senior officers to work with our internal organization of the American Association of Blacks in Energy and other groups. A CEO can lead by example.

SUMMARIES OF MORNING BREAKOUT SESSIONS

Recruitment, Retention, Advancement: What Works?

Workshop participants were assigned to one of four breakout groups and asked to discuss the recruitment, retention, and advancement of women and minorities in engineering careers, focusing on what works, how you can tell if it works, and the characteristics of success. The findings of each discussion were summarized and presented to the group as a whole.

Group 1 Summary

Sandra Begay-Campbell
Sandia National Laboratories

We focused primarily on two areas. First, we had a healthy discussion about the positive and negative aspects of affinity groups, which were highlighted as a best practice. Second, we talked about the image of engineering and what needs to be done to change it.

First, we talked about the successes and failures of affinity groups. We agreed that they provide an internal mechanism for networking and acknowledging work. Affinity groups can raise issues that may not otherwise be heard. On the negative side, some people don't want to separate themselves from the whole by being a part of an affinity group. We had a lengthy discussion about that.

We then turned to a discussion of the image of engineers and keeping young students interested in engineering. We concluded that it is important to convey a positive image of engineers through role modeling and to give young people examples highlighting what engineers do in their careers. Our focus should be on putting students in touch with people who do actual engineering work.

Group 2 Summary

Arnold Allemang
Dow Chemical Company

Our group was able to distill our conversation down to four deliverables. First, when making the business case for diversity, you should provide the data both internally and externally, provide solutions, and then show sustainability. You have to make the best case for diversity. A lot of us who have been engaged in this activity for some period of time feel that the case has already been made and is overwhelming. But there are others who say, I am new to this particular area, and I really need to hear the case made. As a group, we could provide a very good business case for diversity that pulls in data from many different sources. Having participated in diversity activities for a long time, I am of the opinion that the case has been pretty well made, but I can understand that others could have a different point of view.

A Web site called *diversityinc.com* is an excellent source of external data; anybody can contribute information to this site if they are so inclined. Many people may not be aware of this particular source of information, which could provide external data to help you make the case for your company. In most companies, the diversity data is shared internally. But as we heard in our discussion of sharing data here, there are differences in the sharing process. Sharing information about your diversity activities, your diversity programs, the kinds of things you engage in is different from sharing your representation data, that is, the physical makeup of the company. That is a very important distinction. But let's share information about what we do and how that impacts society. We can keep the representation data inside the company.

When our group discussed programs from kindergarten to college to coming to work for a corporation, we found that some really interesting things were happening. One company put up some corporate funding for employees to spend up to 20 hours at an elementary school, a middle school, or a high school. Once the program got under way, a lot of people began spending a lot more time, giving a face to engineering, and money flowed to the school. As someone pointed out, though, every school needs money and we may not be getting it to the right ones. The point is that a lot of people in the company are engaged in that activity, particularly women engineers, African-American engineers, and Hispanic engineers.

Our group also felt that we ought not to talk about "underrepresented minorities" here. We ought to say what we mean. We are talking about women. We are talking about Hispanics. We are talking about African-Americans. If we want to talk about them, we should say so.

To be sure that you can successfully recruit a diverse group of employees, our group argued that you have to start early, and you have to stay engaged. You have to start in the school system by encouraging kids to take an interest in science and math. You have to be there when they are ready to enter the university to offer scholarships and other support. You have to stay engaged during their university years so you have an opportunity to recruit them into the corporation at the end of the day. Once they are in the corporation, it is very important to keep them engaged, inform them of the opportunities for growth in your company, and provide interesting work for them to do. That is a key message there. We are all doing a lot of the same things in terms of the kind of recruiters we use, the kinds of role models we provide, that is, people who look like me recruiting people who look like me, getting executives lined up with universities, and so on.

Finally, I will sum up our conversation about sustainability. We suddenly find ourselves in an economic slowdown—that may be a great understatement for some of us—and we may suddenly begin to lose support in companies for our networks, our African-American network, our Asian network, our Hispanic network, and our women's network. We have to maintain the support of senior leadership for these activities. In these tough times, mentoring becomes even more important. Minorities need connectivity in the organization, not only with senior management, but also with each other, so they can reinforce each other during the tough times, as well as the good times. We have to be willing, and we have to have leadership, to keep these programs active and alive during times of slow economic activity because, historically, they have sometimes been the first to go.

Group 3 Summary

Dundee Holt
National Action Council for Minorities in Engineering

We had a very interesting, very diverse group that did a wonderful job in terms of focusing on what I will call headlines. I am not an engineer. I am a reporter, a journalist, so I think in terms of headlines. In our group, we had representatives of corporations, precollege educators, university educators, a policy "wonk," and not-for-profit groups. We also had representation from Poland and Austria.

Our first headline was how to translate the CEO's commitment throughout the organization. We all agreed that support and commitment are not enough. In other words, a CEO's commitment must filter down through the business units, through the general managers, et cetera. There are a number of ways of doing that.

Cordell Reed (senior vice president, Commonwealth Edison Company, retired, and chair, NAE Committee on Diversity in the Engineering Workforce) described taking top executives on a retreat to acquaint them with the issues. We have discovered that racism is a huge issue for a lot of people. But we think overt racism is much less of a barrier than ignorance of racism, the fact that racism is invisible to a lot of people. They simply don't know. So, we have to raise awareness about the continuing presence and pervasiveness of racism. Professor Klod Kokini (Purdue University) talked about the importance of an emotional pull. Once it is awakened, and we have got to work so that the people at the top feel it, they really do push it down through the organization. The awareness must be institutionalized through policy statements and practices. One suggestion was that diversity targets be linked to compensation, which is already being done at some companies. Motorola is a wonderful model. If a manager there is going to promote people, move people up, the manager must have people from diverse backgrounds in the mix, or explain why. Diversity targets are tied to compensation at Motorola, and people are held accountable. So the first headline is to translate the CEO's commitment throughout the organization.

The second headline, and this also seems to be a "no brainer," is to support scholarship and fellowship programs that reach the target audiences and are not territorial. All of us would like to have our own individual programs. For instance, we would like to know that, if we support a student, that student is going to come to work for us. But, as long as corporations think that way, we will not make the advancements we need. Commonwealth Edison hired a lot of students from Chicago State, rather than from Northwestern or the University of Illinois at Chicago, the schools you see in the literature. But you have to find where the talent is and go there. It's very important that you not be territorial.

We heard about the wonderful programs at Lucent Technologies. One individual, for example, had gone from Lucent Technologies to Motorola and then to Georgia Tech, where he is working on a research program along with NACME. It is very, very important that companies support scholarship and fellowship programs that are going to have an impact not just on individual businesses, but also on the state of the industry. I am very proud of Sandra Begay-Campbell, the product of the NACME Scholars Program, and I'm proud of the corporations that support the program that don't demand that NACME scholars go to work for them. She has done a great job at Sandia National Laboratories and a great job as the head of the American Indian Science and Engineering Society.

We also talked about the importance of mentoring, another very, very important way of supporting scholarship and fellowship programs. We need to change our idea of what constitutes the best and the brightest and where we can find the best and brightest. This is a huge challenge, I think, for all of us. We had a serious conversation about community colleges. Toni Clewell, from the Urban Institute, described the Lewis Stokes Program, an example of the importance of making linkages between community colleges, universities, and corporations. We should celebrate the fact that a large number of minorities and women who have reached high levels of achievement have come from community colleges. Richard Tapia, from Rice University, is a brilliant mathematician who started at a community college. We need to remove the stigma of community colleges. We can do this not only in our own organizations, but also by supporting scholarship and fellowship programs that reach our target audiences, not just enhance our corporate images.

The third headline was another no brainer—that we build stronger links among business, industry, government, and educational institutions. It was interesting that our discussion went on for about two minutes before we realized that we hadn't included educational institutions. We were talking about what business and government could do to improve education but not how better education could help business and government. It is very important that we build those linkages and that we come together as equal partners. It cannot be one handing down edicts to the other. People involved in education mentioned that they could really use industry support, for instance, in lobbying and particularly in support of their bills before Congress. These bills may appear to be simply politics, but they have an impact on industry, and education is the vehicle that makes industry's success possible.

To sum up, our three headlines are: (1) to translate CEO commitment throughout the organization; (2) to support, in a nonterritorial way, scholarship and fellowship programs that reach target audiences where they are; and (3) to build stronger linkages between business, industry, government, and educational institutions.

Group 4 Summary

Gary Downey
Virginia Tech

We had a very focused discussion, maybe the most focused discussion in which I have participated in a breakout session. We came up with categories that mesh with the categories of the other groups. As I was listening to the other summaries, it occurred to me that we are mapping out the features of a report that would be structured in such and such a way. Then I thought, well, no, we are mapping out the site design for a potential Web site. Then I thought, well, wait a minute, maybe we are doing both. The report can't be too long, so we could use certain exemplary best practices, and it could have a bang-up argument with some illustrations and pointers to a data-rich Web site for people to use as a source of further information.

We divided our conversation into three categories: recruitment, retention, and advancement. Under recruitment, we talked about the recruiter's authority, the recruitment process, ongoing efforts, rewards for candidates, best practices, and indirect action. Under recruiter authority, we had one only entry, that the recruiter must have the authority to seize control of a situation and hire strategically. Therefore, recruiters must have certain characteristics. They must all have deep personal commitments. The recruiter team must be diverse. The recruiters must be trained to think about hiring in a way that brings in criteria that maybe were hidden before. Also, rewards and empowerment must be provided for successful recruiters.

What about the recruitment process? Recruiters should look beyond the candidates who have signed up for interviews, consider nontraditional pathways, and develop focused relationships at selected schools. They should make their current diversity initiatives known to attract candidates from wider populations and, finally, apply the principles of diversity across the entire organization. In addition to individual recruiting processes, each organization should structure an ongoing program for mentoring undergraduates and graduate students and employees.

What about the candidates themselves? A company should provide incentives for the candidates, including financial incentives. There might also be actions that may not have immediate benefits but may help over the long term. Precollege-education outreach programs fit into this category.

On to the second large category, retention. First, a company must conduct internal research, for example, to measure the retention of individuals, and then they must share that data. One way to do this is by actively soliciting feedback through the focus groups we heard about this morning. A company should communicate its visions and values, actively survey employees, and, in short, routinely research its own organization. Finally, a company should develop and share metrics for analyzing and following cultural changes, climate changes. If you can put together a reasonable amount of material, you might not be able to bring everyone in the company on board, but you could go a long way in that direction.

In terms of professional development, employers should consider the first five years of a new employee's company career as an extension of the recruiting process. Mentors should be provided from outside of the immediate chain of authority, and the company should provide social activities. Finally, employees should be aware of possible career paths in the organization.

As for formal personnel practices, a company should have zero tolerance for harassment and other unacceptable behavior. Other practices could include providing an ombudsperson to resolve disputes, establishing an "on-boarding" program to help new employees adapt to the company, analyzing job competencies, and having a good system for posting jobs. A company can build visibility through employee recognition programs, for example, and recognize positive contributions. A company can build horizontal networks by strengthening local diversity networking groups inside the company and supporting other kinds of informal programs for building relationships across the organization horizontally. Finally, a company should also build vertical relationships, for example, by encouraging connections between new employees and individuals in upper management.

Our last major category is advancement. In our discussion of leadership development we discovered that mentoring means something different in the first few years than it does later in one's career. Early on, the focus is on gaining admission, getting a sense of fit, of belonging, a sense of connection. Later on, the focus is on building organizational knowledge so that one can advance.

Mentoring outside the group becomes important here. Another good practice involves tours of duty, rotational assignments that develop an employee's knowledge of different parts of the organization. Companies can also use other strategies to build organizational knowledge and make career paths visible. Finally, a company should develop and share its metrics for success.

The very last category, and probably the most important overarching category, involves making the management commitment to diversity visible through programs inside and outside the organization. One example is Motorola's succession-planning program.

Voices from the Field
A Panel Discussion

National Society of Black Engineers

MICHELE LEZAMA
Executive Director
National Society of Black Engineers

As the executive director of the National Society of Black Engineers (NSBE), I was very happy to be invited to speak on this topic. A unique aspect of NSBE is that we are a student-run organization, so I can give you a true perspective on what my students think about diversity in the workplace.

Every March we hold our national convention. For those of you who have not been to an NSBE convention, I invite you to attend. It is a fantastic event, attended by more than 8,000 African-American students who are currently enrolled in engineering curricula. The convention includes roundtables and best-practice discussions, as well as career fairs. It is a great opportunity to see how your workplace practices are perceived by and influence the current student population.

At the national convention, we give the students a survey called the "NSBE 50," asking them where they want to work, why they want to work there, what works for them, why they are thinking about working at this place, and general questions about what they think and the relevant factors in picking a place to work. The NSBE 50 survey, conducted by Bowling Green State University, is statistically significant, and every year we present the results to our Board of Corporate Affiliates, our premier sponsor group.

Last year, the conference was held in Indianapolis, Indiana. A total of 2,480 people participated in the survey, including 26 percent of the convention attendees and 23 percent of the NSBE membership. Responses were generally representative of the entire NSBE student membership in terms of gender, age distribution, et cetera.

The overall question was for what companies in America would students most like to work. Table 1 shows the results for 2001. Respondents were also asked why they did not want to work for their last choice (Table 2). Atmosphere,

TABLE 1 2001 NSBE Top 50 Employers

1. IBM Corporation	18. The Boeing Company	35. **Eli Lilly & Company** (+)
2. Microsoft	19. AT&T	36. **Kraft Foods** (+)
3. General Motors	20. Sony Electronics	37. Nike
4. Motorola	21. ExxonMobil	38. Delphi Automotive
5. Accenture	22. Goldman, Sachs (+)	39. DaimlerChrysler
6. Ford Motor Company	23. Verizon	40. FBI
7. General Electric	24. Corning (+)	41. Abbott Laboratories
8. Intel	25. Dell Computers (+)	42. **Agilent Technologies** (+)
9. Johnson & Johnson	26. Disney/Imagineering	43. **Bell South** (+)
10. Lucent Technologies	27. Dow Chemical Company	44. **General Mills**(+)
11. NASA	28. 3M Company	45. **3Com Corporation** (+)
12. Lockheed Martin	29. Raytheon	46. Merrill Lynch
13. Cisco Systems	30. Nortel Networks	47. Kimberly-Clark
14. Merck & Company	31. **Honda of America** (+)	48. CIA
15. Hewlett-Packard	32. Du Pont Company	49. **Texaco** (+)
16. Procter & Gamble	33. Coca Cola	50. **Medtronic** (+)
17. Texas Instruments	34. Nokia	

Bold = new to the NSBE top 50 employers. (+) = moved up at least 10 places since last year.
Source: Bachiosi, P. 2001. NSBE 50 2001: mix of optimism and caution. NSBE Magazine 13(1):33.

corporate culture, and diversity were identified as the fourth most important criterion for why students would not want to work at a particular company. Diversity is truly relevant to students' decisions.

Companies identified in the NSBE top 50 rated highly on campus visibility, marketplace visibility, diversity programs, and concern for the environment. Clearly, diversity is on the minds of students thinking about where they want to work. Salary becomes a more significant factor for students as they get older. It was more important to seniors than to freshmen and sophomores. Underclassmen were more concerned about the recruiter's technical knowledge, name recognition, and other visibility factors.

The significant factors for choosing a particular company are shown in Figure 1. The importance of diversity programs remained constant from 2000 to 2001; an average of 80.5 percent of respondents rated diversity programs as very important or extremely important. Advancement opportunities was the most important factor, followed by job security, interesting work, training opportunities, and work/life balance. The diversity program was one of the six most important factors.

The relationship with the black community is also very important, not so much the race of the recruiter, but how much the company is involved in the community and how many blacks are in management, in higher level, visible positions, not just in entry level positions.

TABLE 2 Why Students Do Not Want to Work for Certain Employers

Type of industry/product; consistent with interests/major; interesting work; type of work	21%*
Knowledge about company; like the company; quality of product	14%
Government/military/security clearance	10%
Atmosphere; corporate culture; diversity	7%
Stability/job security; success of company	7%
Reputation; visibility; size of company	3%
Career advancement/development; personal growth	2%
Impressions based on a company presenter or recruiter	2%
Accidents/safety	1%
Diversified company; innovation/technology	1%
Moral issues	1%
Personal experience; work(ed) there; internship/co-op	1%
Salary; benefits; location	1%
Miscellaneous	4%

*NOTE: Percentages of those giving a reason. 27 percent did not give a reason.

Later on in the survey, we asked about the importance of recruiting criteria. Where should companies focus to improve their recruiting? The results of the survey (Figure 2) indicate that companies should focus more on the recruiter's knowledge of the position to be filled. Only 28 percent of the respondents rated the race of the recruiter as very important or extremely important, whereas 88 percent rated position knowledge as very important or extremely important. When you send a recruiter out to recruit a minority candidate, the recruiter should have specific knowledge of the position. The black students at the NSBE conference felt that knowledge of the position, what they would experience when they got to the workplace, what type of job they would have, and what their day-to-day functions would be was a lot more important than the skin color of the person talking to them. They really wanted to speak to someone who knew a lot about the job.

Perceptions of importance vary from freshmen to seniors, and the survey compared freshman and senior responses. As a student advances through the college program, concerns about where they will work and the reasons for selecting an employer change. Factors that become increasingly important are: the number of blacks in management; innovation; diversity programs; interesting and challenging work; and response time after the interview. Factors that become less important include: co-ops and internships (for obvious reasons), and scholarship programs, which become less important as students approach graduation.

Every year, we have a special category of issues relevant for that particular year. This year's topic was racism, affirmative action compared to diversity plans. Our recruiting companies want to know if it makes a difference to a black student whether their recruiting materials focus on affirmative action, using

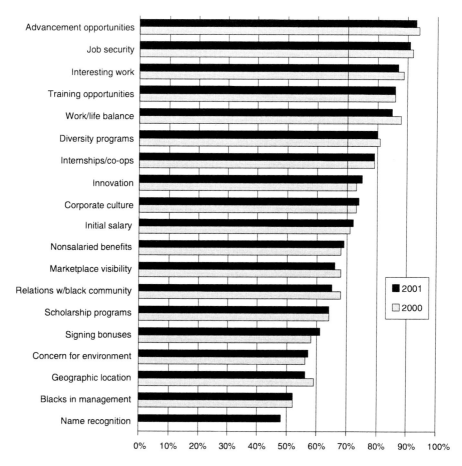

FIGURE 1 Employer selection criteria. Percentage indicating very important or extremely important.

those words, or on diversity. The survey results indicated that the students don't differentiate between affirmative action programs and diversity programs. However, the survey responses indicated that the students did view preferential treatment and special training programs negatively. If a company has a program just for black students or just for minority students, the students view it as a negative.

Based on the survey, we make recommendations to our corporate partners about what they should consider in their recruitment practices to attract our students. Company visibility is key, getting the company name out there, as well as information about the company's involvement in the black community. The students suggested that companies do this via participation in career fairs. Other mechanisms for increasing company visibility include sponsoring organizations

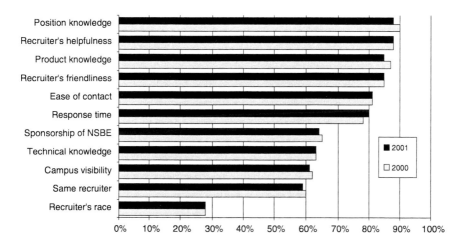

FIGURE 2 Recruiting criteria. Percentage indicating very important or extremely important.

relevant to students' needs, keeping the company Web page current, and contacting students directly by e-mail.

Companies should recognize the importance of a recruiter's characteristics, such as accessibility, responsiveness, and knowledge. Knowledge was considered the most important throughout the survey, not the color of the recruiter's skin but how much he or she knew about the company and the experiences the students would have in the workplace. Students were looking for diverse opportunities in the workplace rather than diversity per se. They want to know if they're going to be stuck in the same position or if they'll have the opportunity to learn and grow once they are in the workplace. Internships and co-op experiences are also effective recruiting tools. Overall, NSBE students indicated that, if they had a chance to work at a company and experience it for themselves before graduation, they would be more likely to want to work there.

We also asked students whether the Internet was an effective recruiting tool and if face-to-face relationships were still necessary. The students felt that the Internet was definitely helpful in recruiting, but that face-to-face interaction is still very important. The last point that stands out in the survey results is that companies should avoid using language that suggests preferential treatment in recruiting materials. Students view preferential treatment based on race negatively.

We publish complete survey results in a national magazine, called *NSBE Magazine*, which is published five times a year, every two months during the school year. We also publish a high school magazine, called the *NSBE Bridge*, to encourage minority youth interested in math and science to go into engineering. These publications are good vehicles for getting your message directly to students.

Society of Women Engineers

SHELLEY A.M. WOLFF
President
Society of Women Engineers

The Society of Women Engineers (SWE) is an advocate for diversity. We embrace this commitment in the opening line of our diversity statement, "SWE acknowledges and respects the value of a diverse community." We firmly believe that every engineering and technology organization has a responsibility to ensure equity for women and minorities at all stages of their careers—from entry level to CEO. One of SWE's missions is to encourage women to achieve their full potential in their careers, which can be a challenge because they sometimes feel isolated and out of the loop in their companies.

Based on talking to our members and my own experience, I recommend three things that you, as corporate leaders, can do to encourage diversity, participation by women in particular, in your company: (1) support women's groups, both internal and external to your company; (2) highlight and advertise the achievements of women in your organization; and (3) ensure fairness and equity in promotions at all levels. Let's take a look at each of these three items.

The first item, supporting women's groups, means providing the means for women to network. A network provides mutual support, as well as a forum for women to exchange information and ideas. Many companies, especially large companies, support formal women's programs in the company, providing workshops and speakers. In the Kansas City office of my company, HNTB, we have a much smaller group of women that operates on a much more informal basis. We go out for lunch every couple of months to catch up on what's happening in our lives and careers and in the company.

Both of these models, formal and informal, help women connect and support each other, which is essential to affirming our value to the company. A women's group is a forum where we can celebrate our successes, large and small. I also firmly believe that companies should encourage and support involvement by women in external minority organizations. These are especially valuable if there is no internal support structure in your company. Too many firms sponsor an employee for only one professional organization, usually a technical society. This policy should be expanded to include a second professional association, such as SWE, the National Society of Black Engineers, the Society of Hispanic Professional Engineers, or the American Indian Science and Engineering Society. SWE members relate many instances of being able to obtain company support to attend their technical association conference, but not the SWE annual conference, which provides technical, management, and leadership training opportunities. This attitude must change. Companies must invest in personal development, as well as technical development.

The second thing you can do as a corporate leader is to highlight the achievements of women in your organization. My company publishes a wonderful marketing-type magazine for clients and employees and a newspaper-type publication for internal use. When major projects and promotions are reported, photos of those involved are included. As you can imagine, in my 21 years with HNTB it was very discouraging to see photo after photo of white males only. Our director of corporate communication finally recognized this problem and began to seek out women who excelled but had been overlooked:

- a woman environmental engineer in our Indianapolis office who is leading a megaproject, a complex, multidisciplinary project with high visibility in the community and more than $5 million in fees
- a woman engineer in Louisville who is not only a top-performing project manager, but also a teacher of undergraduate courses at the local university
- a woman interior designer who is responsible for our continued success with a major client, not always for large projects, but the relationship of trust she has built with the client ensures continuous work and revenue for her department

Women's achievements are often overlooked simply because their supervisors or line managers do not champion their efforts. The third thing you can do is to ensure fairness and equity in promotions. Women will leave your company (often to start successful businesses of their own), if they do not feel your company's job assignments and promotions are fair. Too often we see men promoted to positions based on their potential and who is sponsoring them, while women must already be performing at the higher level before they are deemed qualified for the position. We must stop this inequity.

My company has initiated a succession-planning program that I feel will help address this issue. Detailed job descriptions and required skill sets will be posted for high-level positions, such as group directors and office and division leaders. Employees will be evaluated for their readiness to move into these positions, and action plans will be prepared for those who are not ready, outlining the job assignments and training they need to compete for promotions. This type of program is not a cure-all, but it could open a lot of doors to women in our company.

In conclusion, to remain competitive, to hire and retain the best talent who can find the best solutions to engineering problems, you must train, encourage, and champion minority candidates, including women. Women provide a wonderful reservoir of engineering talent that is not being fully used. Taking advantage of it will help you build for the future.

Society of Hispanic Professional Engineers

ORLANDO A. GUTIERREZ
Past President
Society of Hispanic Professional Engineers

I represent the Society of Hispanic Professional Engineers (SHPE), of which I was president from 1993 to 1995. I am still associated with SHPE, and my feeling is that if it didn't exist, somebody would have to invent it. Hispanics are probably the least represented minority in the workplace. They are the least represented minority in government employment and engineering and science on all levels. This morning, one of the speakers stated that the Hispanic population in engineering has increased to 4.8 percent. Compare that with 13.9 percent Hispanic representation in the population as a whole right now, and more than 20 percent representation in the nation among individuals under 20 years of age, the next generation coming out of college. There is obviously something wrong, but who is to blame? I think the Hispanic population is to blame. I think government organizations are to blame. I think American business is to blame. I think the educational and academic institutions are to blame. There is plenty of blame to go around.

SHPE was founded in 1974 by seven engineers working for the city of Los Angeles who found themselves in a poor working environment. Since then, the organization has grown to include more than 1,400 professionals in science, engineering, and technology and more than 7,000 students in those fields throughout the country. We have 43 chapters from Seattle to Florida, from Maine to San Diego. We have chapters in about 160 schools. We are comparable to an NSBE for Hispanics, but we are a little bit different.

When a company wants to recruit from underrepresented groups, it should not just consider minorities in general, but should look at the particular group they are interested in. Once that decision is made, the company should find them and treat them appropriately.

Our organization is based on a process of self-help, our people helping our people improve at the college level, at the work level, at the precollege level.

The main tool we use is very simple—networking, networking at all levels—professionals networking with college students; advanced college students networking with incoming students to increase retention, college students mentoring precollege students. In order to do that, we have a few main programs. One is a scholarship program. Another is career fairs for employment; our career fair is benchmarking against NSBE's. We don't have 7,000 people there, but we now have 4,000, so we are creeping up. Our career fair is perhaps the best source of Hispanic undergraduate potential employees for companies. Our students are spread among colleges and universities that may have 3 or 4 percent Hispanic students, which means if you go to one university for one day of recruiting, your recruiter may see 16 people, 4 percent of whom will be Hispanic. I think that represents half of a person. The chance of that half person fulfilling your requirements is practically nil. By participating in the SHPE conference, you can see 3,000 students in one place and can probably find somebody to fit your needs, your characteristics, and your interests.

We also run a lot of outreach programs in which our college students tutor and mentor high school and junior high students, getting them into the engineering/science mind frame. If we wait until students are seniors in high school or at the bridge level, it's already too late. Interest in these careers has to be generated when students start junior high, which is the most significant point in a student's life. We also have a leadership training program to teach skills that are not taught at the university, such as how to operate a program, how to write grants, how to run an organization. We also produce a magazine, *SHPE*, which is one of the largest, if not the largest, Hispanic engineering magazine in the country.

So what? So, we have programs. Are these sufficient? No. Are they effective? We have made some improvements. We have affected the retention level of freshmen in college by associating them with successful upperclassmen. Have we been able to help companies in recruiting? Yes. Companies that attend our conference and show their flags and show real interest, which is difficult to fake, are considered by our students to be employers of choice. Are we satisfied with what we are doing? No. A project like this—to help the community, which helps industry, which helps the nation—requires money. We don't have enough. We have a scholarship program that gives about $275,000 a year in scholarships. That is a pittance. We are able to help about 270 students a year, but we have a list of 1,400 applicants, all of them worthwhile candidates.

We spend about $120,000 a year for our outreach programs. Is that sufficient? It is a drop in the bucket. It should be 1.5 orders of magnitude bigger than it is. Can we generate the funds ourselves? No. We need to partner with industry, not to ask for gifts, but to seek investments in the future. You won't see the results tomorrow. You won't even be able to measure them in a couple of years. But you will see results in the future. I think that support of organizations like ours, like NSBE, like SWE, is one of the best investments American industry and government can make.

American Indian Science and Engineering Society

SANDRA BEGAY-CAMPBELL
Executive Director Emerita
American Indian Science and Engineering Society

The American Indian Science and Engineering Society (AISES) has a lot of similarities with the other programs presented here, such as accessibility to students, a magazine, scholarship support, and support from corporations. But as an organization that is mostly student-based, we focus a good deal on American Indian communities. Part of our approach is to instill students as soon as they arrive in college with the idea that they are unique, that they have the potential to be leaders in their communities because there are so few of them in the university system, and that they have an obligation to show leadership, as well as to give back to their communities. Our focus is on increasing the population of American Indians who want to go into science and engineering careers. I am a product of that approach.

Today, I will tell you about my personal experience, as well as about my experience organizing, fund raising, and talking to corporations and government agencies on behalf of AISES. My perspective is based on my experience running the organization for two years. I have been with the AISES organization since 1983, so I understand it from the grass roots up, from a student perspective as well as from a leadership perspective. The unique thing about AISES is that it represents a diversity among American Indians. Most people think there are typical characteristics of American Indians, and there may be a few. But with more than 500 different tribes in the United States, there are also great differences. Knowing that from the onset will help you avoid saying, no, I thought American Indians were like this or like that. Mostly, people's ideas are based on old Western movies. It is a little sad to know that, even in this day and age, people have a perspective of American Indians from television and the movies, but that is the truth of it.

I like to tell a joke about one of our students who is of the Mohican nation out of New York state. He gave a little talk about who he was at a student forum, and a fellow American Indian student got up and said, "God, I thought all of you guys were dead." Where do you think he got that? From "The Last of the Mohicans." Even the students themselves have some misperceptions about American Indian culture. So, a lot of educating needs to be done for people to understand the issues.

As you go from region to region, you find some similarities among the tribes, and I will highlight some commonalities that relate to recruitment and retention. I can't talk too much about advancement because I haven't found an American Indian scientist or engineer in an executive or top management position yet. I am still looking though, and when we track that person down, we will certainly highlight his or her accomplishments. For now, I will talk about role modeling instead of advancement.

Because AISES has been involved in these issues since the eighties, I have seen a lot of recruiting strategies that have worked. One of the keys to success is consistency. If students see you or someone from your company as a recruiter, and they see you year after year, they begin to know who you are. That is how I was recruited for every job I have had. I knew the recruiters who attended those meetings, and they were very cordial. At first, I might not have been interested in their companies, but they took the time to find out about me. And when I was ready to be hired, I found some time to find out about them.

That is a principle we instill in our students. Know who you are going to work for. Find out how they will support you once you get into the company because retention is a big, big challenge. Consistency and getting to know our students are big factors in successful recruiting. After trying to nurture professionals and watching a handful out there practicing and working since the early eighties, I've learned that it takes a long time to "grow" a person like myself. We are very careful about who we allow to recruit our kids, and we take a very maternalistic, maybe paternalistic, role.

Because there are so few American Indian students, we want them to have the best opportunities. We try to teach them to decide what they want, where they are going, what type of work they will do on the job site, but also to make sure that there will be a nurturing environment so they won't get lost or feel left out. In truth, they are the minority of minorities, and that is a fact they have to live with. And we tell them that. We are very frank with them. For example, a student might be the only American Indian out of 8,000 employees. He or she might have to be an advocate for American Indian issues, answering the goofy questions that come up, like "The Last of the Mohicans."

We prepare our students for the career fairs by informing them where companies are coming from to offer them jobs. We also educate companies to be mindful that when they hire an AISES student they should be aware that the pool is small. The newly hired AISES student will then have an obligation to try to inspire another American Indian young person to follow in this path of education. We deliberately instill the notion of giving back to the community, that once a person is successful and has secured a nice job, he or she has an obligation to find a way to support another student.

When the students come to an interview, they will have a lot of questions. How can they be a part of the community? What kind of outreach programs do you have? Can they participate in community projects? It may be a mistake in how I have managed my own career, but I have made it clear that I consider this a part of my job. I have told prospective employers I won't be able to do the best job for their companies otherwise. I have to give back to the community. If you hear that common theme, you will recognize it as something we've worked very hard to instill in these students.

When you talk about the retention of American Indians, you must remember that these students have left their homes and their families, that they may have lived their whole lives in rural, isolated circumstances, sometimes on a reservation. Remember, they need to maintain a family connection. We also encourage affinity groups and networking, which can help them feel they are a part of a group.

Two companies in particular have done this extremely well—both Lucent Technologies, which has the Lucent United Native Americans (LUNA) Group, and IBM Corporation, which also has a very strong affinity group. These two groups actually have conventions at different places in the country to bring Native American employees together once a year. That is a very innovative idea. At Sandia Labs, where I come from, there are 200 of us, and we are a pretty tight-knit group; that is an important factor in retention.

Another aspect of managing diversity is promoting minority employees who have done well or who are doing outstanding jobs. In some American Indian tribes, it is not a natural tendency to boast about your accomplishments. It is a learned skill. We try to remind students that for people to acknowledge your work, you have to talk about yourself, which is sometimes difficult. You don't have to boast about your own work, but you should be able to talk about your accomplishments. To be promoted, your job evaluation will be looked at to see how well you have done. The students have to know that that is something to expect, so coaching them can be very helpful.

If you can help them understand the culture of your company, it will help them understand how they can participate. On the other side of the coin, you also have to do your homework because there are things you should understand about their culture, like which tribe they belong to. They are not just American Indians—each individual belongs to a specific tribe. They may want to travel back to their homes for certain occasions, for example.

When we talk about advancement, the only thing I can highlight is, on a national front, that some American Indians are pushing the bounds of science and engineering. You should know who they are. In fact, one American Indian engineer, John Herrington, will be going up into space in August of 2002. He will be the first Native American to go into space. You should know his name because he is the only one. We have deliberately included him on our board of directors, where he can become a visible role model and mentor with a platform from which to talk about his work and the issues facing American Indian students.

Part of my job as executive director of AISES was to talk about issues in the workplace to make sure students understand the decisions before them. When we talk about role models and mentors, we are not talking about them in a generic sense. We want them to share real experiences and real difficulties. Take my friend who worked at IBM who had very long hair. He lost a relative. In his tribe when you lose a relative, you cut your hair short. The talk among his colleagues was, oh, you cut your hair to conform to IBM, when, in fact, he had made a very personal choice to show he was in mourning for the loss of a family member. So, you see, some of the issues students have to deal with have to do with their tribal cultures. They have to balance them with doing the work they have been hired to do.

The last thing I will leave you with is to tell you there is a huge national network for the groups represented on this panel to work together. Over the years, we have tried to keep dialogues going by participating in panels like this where we can talk about our differences. But there are also some nice common threads running through all of these organizations. We are here to help you, as representatives of industries and universities, understand more about the minority and women students and professionals who want to enter the workplace.

Question and Answer Session

Dundee Holt (NACME): Sandra, the information you shared with us, particularly about promoting oneself, which is alien to a lot of American Indians, and the fact that you couldn't even identify one American Indian who is a high-level executive, raises some questions. How does that conversation between a young American Indian engineer and his or her manager take place? How does AISES encourage students or new employees to have that conversation, to raise those issues?

Sandra Begay-Campbell (AISES): We tell students that when they speak up, they are not doing it just for themselves, but for all people like them. The students have to be aware that people are pushing the bounds of participation at a national level, where you have to have a voice. This puts a burden on the individuals who are speaking up at that level. If I hadn't shown up for this meeting, there might not be a voice for AISES or American Indian issues. There is a lot of pressure on those of us who are trying to raise awareness of the issues. First, you have to show up, and then you have to speak up. You can't just be quiet and listen. You have to say something. Otherwise the moment will pass and you will have lost a chance to bring those issues forward.

Cordell Reed (NAE Committee on Diversity in the Engineering Workforce): Michele, I was struck by the finding on your survey that students do not want preferential treatment, and I am wondering about the issue of mentoring. If there are mentoring programs in corporations only for African-Americans and Hispanics and women, do you think they are viewed as positive or negative?

Michele Lezama (NSBE): Mentoring is a key component of any workplace and of all of our practices in NSBE. We believe in putting real live examples of black engineers who are succeeding in the workplace in front of our students. The finding about preferential treatment was related to how majority workers view them in the workplace. If I was recruited because of the color of my skin and not because of my accomplishments, for example, because there is a well known preferential recruiting program, that makes it difficult for me to succeed. Even if I perform well, some people assume that I was hired because of an affirmative action or diversity program. I didn't intend to say that mentoring is not important. Mentoring is key to any relationship with a minority candidate, particularly in engineering and sciences.

Arnold Kee (American Association of Community Colleges): I'd like to follow up on that thought. When I first saw that reference to not wanting preferential treatment, I was trying to figure out whether you were saying that, if a company views affirmative action as preferential treatment, then the students see it as negative, or that, if the company says it has an affirmative action program, students will be afraid that other people in the company will think they are receiving preferential treatment. As you know, people who oppose affirmative action call it preferential treatment.

Michele Lezama: Right. I encourage all of you to read the full results of the survey in the NSBE magazine. The point is that you have to showcase not only a person, but also the quality of that person's work. It is not just about the color of my skin. It is about the fact that I have three degrees, that I graduated magna cum laude, that I bring a lot to the table. So, don't just showcase me as a "diversity hire." You have hired a qualified individual, and that is what our organizations are putting out, qualified individuals. You have to showcase that as well.

Bill Friend (NAE Committee on Diversity in the Engineering Workforce): First of all, it is very good to see all of you together. My question is if you should be more integrated as a single organization. You have common causes, and I wonder if you would be more effective if you were somehow at least a confederation of organizations.

Shelley Wolff (Society of Women Engineers): We do all have very similar mission statements. We also have specialty areas. SWE, for example, tries to encompass not only K through 12 outreach, but also college students who are the majority of our members, similar to NSBE. But we also follow members throughout their careers. So, we have a lot of similarities, but we also have some areas that we focus on as individual groups.

Michele Lezama: We already work together on a number of initiatives. Gina Ryan, executive director of the SWE, and I work together closely. Jose Rivera from SHPE and I work together. I haven't worked with anyone from AISES yet, but I look forward to that. So, we do talk and we do work together.

I am asked a lot about creating a consortium of minority and women's engineering organizations. But our strength is in bringing together students on common ground based on their cultures and based on their experiences. They don't have to explain, for once, who they are and what they are about. That is really the foundation of our organization, the fact that students from similar backgrounds can come together to work on a common cause, which is excelling in an engineering curriculum. That is what makes our organization so vibrant. If you take that cultural component away and merge all of these groups, you would lose the one thing that makes us strong. At the same time, we have to work together on common initiatives to make sure that diversity is respected in the workplace. But we still need to maintain our strong cultural components.

Orlando Gutierrez (SHPE): If I may add my two cents, I think we agree that each group has its own problems that have to be targeted individually. But there are a lot of problems to deal with, and a federation would be nice. As a matter of fact, I am sorry to say that there is more than one engineering organization in the Hispanic community. There is my organization (SHPE), the Mexican American Engineering and Science Society (MAESS), and the Society for the Advancement of Chicanos and Native Americans in Science (SACNAS), which is a mixture of Hispanics and American Indians.

We can work together, but in the Hispanic community, some people want to focus more on one problem area than others. It is the same with engineering organizations. They are all engineers, but would IEEE and ASME merge into one group? I don't think so. But they should work together, and they do in the American Association of Engineering Societies. I think that concept is very good.

Sandra Begay-Campbell: I want to highlight a special aspect of American Indians, which has to do with reservations and the state of their economies. We hear a little bit about the digital divide in urban areas, and maybe in some rural Hispanic and rural black areas. But Native Americans may not even have electricity, or even the potential of having electricity in the short term. We are talking about some very basic needs that haven't been met yet. We are trying to encourage them to become engineers and to surf the Internet, when they may not have electricity in their homes.

That very particular problem has to be dealt with on a tribal level. To pursue a very technical degree and to keep up with the digital age is very difficult for our students, and the divide is getting bigger and bigger. Some unique problems for each group have to be addressed by each group, but also with partners that are willing to help.

Karl Pister (University of California): I want to pick up on Bill Friend's remark about a consortium. In one state, in several perhaps, the word "minority" has taken on a new meaning. Rather than a consortium of groups that have been called minority groups, we must recognize that they will be (in some places are already) the majority. In fact, they are the majority of the electorate in California.

Why do we deal with the symptoms of the problem rather than the root causes? These organizations are not to blame. We, the people, are to blame, we and our elected representatives. That is the problem we ought to be addressing.

Afternoon Keynote Address

Advancing Minorities in Science and Engineering Careers

WILLIE PEARSON, JR.
Chair, School of History, Technology, and Society
Georgia Institute of Technology

I have been asked to talk about a study I conducted between 1993 and 1995 on African-American Ph.D. chemists. At that time, about 400 or so African-Americans had received Ph.D.s in chemistry. In this study, a random sample of 45 chemists was selected for interviews. All were U.S. citizens who had received their Ph.D.s prior to 1994. Even today we have very scant information about the experiences of ethnic and racial minorities in the scientific and technical workplace. This is the focus of my presentation today.

The study, which was conducted as a cohort analysis, parallels a study I conducted earlier that was published in my first book, *Black Scientists, White Society and Colorless Science: A Study of Universalism in America*, (Associated Faculty Press, Inc., 1985). The first study focused on career patterns of both African-American and Caucasian Ph.D. scientists in the United States. In both studies, the cohorts parallel various civil rights and legislative issues. Cohort 1 received their doctorates prior to 1955; Cohort 2 between 1955 and 1964; Cohort 3 between 1965 and 1974; Cohort 4 between 1975 and 1984; and Cohort 5 between 1985 and 1994.

I will summarize some basic trends before looking at the cohorts individually, lest you think that what happened in Cohort 1 is no longer relevant to more recent cohorts. Prior to the mid-1960s, black Ph.D. chemists were employed almost exclusively in historically black colleges and universities (HBCUs), regardless of the institutions from which they earned their Ph.D.s. The second major sector of employment prior to the mid-1960s was government. As we will see later, this has changed somewhat over time.

The passage of the Civil Rights Act of 1964 brought about a sea change in the opportunity structure for African-Americans in terms of their participation in sectors of the economy from which they had previously been excluded or restricted. Only with the introduction of federal legislation, did new opportunities appear. As a result, Cohort 3 is the largest cohort in the study. In terms of academic employment, only during this period were opportunity structures about equal in HBCUs, predominantly white colleges and universities, and government. Some of the people who were hired by predominantly white institutions were recruited from HBCUs.

For Cohort 1 (before 1955), the opportunity structure for employment outside of HBCUs was extremely limited. A few people were employed in industry and predominantly white institutions; however, the vast majority were employed in black colleges. In some instances, very fair-skinned African-Americans "passed" in order to work outside of HBCUs. However, some individuals who could pass chose not to do so. For example, one fair-skinned interviewee related this experience. An interviewer once asked if he was Indian or Hawaiian. When he replied that he was a Negro, the interviewer's response was that that was "too bad because it is illegal . . . to hire Negroes in our labs." If his answer had been Indian or Hawaiian, he might have been hired. Ultimately, he got a job in industry, where he was very successful. Ironically, the company that first turned him down ended up being taken over by his company. Because of that one African-American, the company had to change all of its meeting sites because hotels at that time (especially in the South) were racially segregated.

In Cohort 1, a number of African-Americans graduated from the top chemistry departments. As many pointed out, they were good enough to be educated in these institutions but not to be hired on their faculties. In one case, an interviewee was the only doctoral student to pass his preliminary exams in a very prestigious department. Nevertheless, he could not be hired on that particular faculty. One of his classmates, who had not performed as well, was hired on this faculty. According to most of the Cohort 1 respondents, this experience was typical for African-Americans who attended these schools at that time.

With Cohort 2 (1955–1964), we begin to see the results of federal regulations promulgated in the late 1950s and early 1960s. One respondent began his career in industry with a master's degree. A white technician refused to work for him, so he had to do most of his lab work himself. Because of the conflict, he decided to return to graduate school to pursue his Ph.D. Upon his return to the company in the mid-1960s, there had been a major change in how he was received. He credited federal intervention (the Civil Rights Act of 1964) for the changes in his company. In this case, he was able to get a high-level position because he fit the company's requirements—a fair-skinned African-American. According to the respondent, if the company was going to have a black person in the lab, that person would to have to look as much like his white counterparts as possible. In his case, he could "pass." However, another problem he encountered was that, being single, the company exerted a lot of pressure on him to get married.

Another Cohort 2 graduate of a top program had a very good offer from one of the larger chemical companies. However, because there were so few African-Americans with Ph.D.s in chemistry, he decided to pursue a career in academia where he could dedicate his career to increasing the number of African-Americans in chemistry.

Cohort 3 (1965–1974) was the largest cohort in the study, primarily because of antidiscrimination legislation and legal challenges to discriminatory practices. Companies and schools that violated antidiscriminatory regulations could lose their federal contracts. Things began to change for Cohort 3, as more black people were able to pursue careers in industry. By the time of Cohort 4 (1975–1984), the vast majority of interviewees were pursuing careers in predominantly white colleges and universities and industry, with government a distant third. By the time of Cohort 4, respondents who took academic positions were most likely to pursue careers in institutions that were very similar, in terms of racial composition, to their undergraduate programs.

One young interviewee in Cohort 5 (1985–1994) explained that he received a number of invitations for job interviews but very few offers. The experience revealed that the belief that blacks with Ph.D.s in chemistry from very good schools had boundless opportunities was indeed a myth. The respondent was invited to apply for a faculty position. The process went on for months, and he felt that he was invited to campus only to meet an EEO requirement. Almost nine months later, he received an offer, which he refused because of the way he had been treated.

Several issues emerged regarding academe. First, regardless of academic affiliation, most interviewees reported they had difficulty securing research funding, and they believed that funds were not distributed equitably. Some respondents at government funding agencies reported that nonblack scientists submitted proposals that were not as well thought out but were funded based upon where the investigators were employed.

Because many respondents saw limited opportunities in major research universities, they chose to work in industry over academe. Although some decided to pursue careers in industry for financial reasons, this was not the only consideration. Others chose to go into industry to increase black representation.

Blacks were far more likely to accept employment in companies they characterized as "welcoming." This was reflected in a comment by one of the highest ranking African-Americans in the chemical industry. While being recruited at the company he chose, people in human resources were very friendly and welcoming. In fact, his hosts took him to meet community leaders in both the majority and minority communities, nonprofits, etc. In contrast, during a visit to another company, it was very clear that the host was a bigot. On that visit, he was shown the most dilapidated housing and the most segregated areas of the community.

Another issue is the climate of the particular unit in the company where the individual would be working. In one company, African-American Ph.D. chemists in different units had vastly different experiences. One was struggling to confront numerous challenges and barriers. This individual had been in the same position for about 15 years. Another was in a very nurturing, welcoming environment and was flourishing. This person reported that promotions and assignments were very fair.

Some respondents felt that high-performing African-American chemists and other scientists were not recognized for their contributions. One interviewee in industry commented about his employer: "I think the administration is slower to recognize and to promote black Ph.D. scientists." Several interviewees said they believed that their contributions were recognized much more slowly than other people's contributions. Many argued that the lack of recognition translated into loss of income—from which they may never recover. Respondents also noted that even supportive environments were not static. They cited cases in which the environment had been friendly or welcoming at one time but had changed when management changed.

One respondent who held a senior position in a chemical company indicated that his company tended to recruit only at select institutions where the management and the faculty had very close relationships. The respondent said, "The fact that I am the only African-American at a senior rank is due to the way the company recruits. We tend not to develop networks that would identify African-American Ph.D. chemists for recruitment."

Many respondents took jobs in industry with the hope of staying on the research and development track and moving up through the ranks. Some respondents indicated that once they attained a certain level on the promotional ladder, they could go no higher. A number of respondents who perceived that they could not reach the level of vice president of research decided to move into management. Although some were able to move into line management, most ended up in staff positions, such as equal opportunity and community relations jobs. These jobs were not on the track for upward mobility, because they were support positions rather than positions that contributed directly to the company's bottom line.

One interviewee said that he had started at a company with tremendous hopes and motivation, believing he would be able to do significant scientific work and win outstanding scientific prizes. Over time, his optimism lessened, and he finally was encouraged to leave the bench altogether and ended up in a management position.

Turning briefly to government, we found that a number of people in the sample who had worked at various federal agencies with missions related to science and technology had very different experiences. Although some had done quite well, none was promoted to a senior rank. In one case, an individual who held a high-ranking position in a federal agency had applied for the position earlier but had been passed over in favor of a colleague. In the end, the colleague did not perform well and was fired, allegedly for incompetence. Under new management, the respondent was promoted. In his new position, he had access to his own personnel records, where he learned that his previous opportunities for promotion to the senior rank had been denied because he was not considered "technical" enough. He also discovered that his critics had earned their Ph.D.s from the same university as he had and had taken similar classes.

All of the women in the study were employed in industry or academe. Most reported experiences similar to those of other women scientists. For women, gender mattered as much, or more than, race. Some of the women noted that chemistry in general is a very sexist field and that this was also true in academe—regardless of the racial composition of the institution. They also indicated that they were not included in the so-called invisible college or the good-old-boys network. Some women respondents reported that when they attended professional meetings, especially meetings of the American Chemical Society, their questions were not taken seriously and, in some cases, were even dismissed. Women respondents also reported being given heavier teaching loads and less respect from male students than their male counterparts.

A fairly large number of interviewees indicated that race still affects opportunity, which has also been shown in other studies of African-American professionals. A few respondents indicated that the role of race in their careers had been insignificant. Although these respondents still believed that race matters, they also believed that, in the end, good science would be recognized.

Most respondents believed that industry provides the best opportunities for career advancement for African-American scientists. Most were not optimistic about advancement in major research universities. More than 90 percent of respondents indicated that if they had to select a career again, they would choose chemistry because of their love for the discipline. Nevertheless, they recognized that the chemistry community needs to be changed.

Finally, most of the women (and some men) reported that they would tell young women considering a career in chemistry the truth about the discrimination they are likely to encounter. Armed with this information, the respondents believed that new black chemists will be better prepared to survive and thrive.

Questions and Answers

Suzanne Brainard (University of Washington): Do you think the findings of your research on chemists would transfer to the field of engineering? Is anyone doing similar work focused on engineering?

Willie Pearson, Jr.: Some of my earlier work included engineers, as well as other scientists, but no one in my area, the sociology of science and the history of science, is focusing on engineering. Some people outside the field are looking into these things, but their methodology is problematic. Joyce Tang, for instance, is doing some work on Asian-Americans in engineering, but her work has been criticized for using old databases for engineering and some sciences that are fraught with policy issues. If you don't understand the policy context, your interpretations of the data can be critically flawed. One thing that probably should be done, and I think NACME has tried to do this, is an in-depth study of minority careers in engineering. But scholarship in this area is not at the same level as it is in science. Even in science, most studies, at least until very recently, have focused on academic careers rather than on careers in industry.

AFTERNOON PRESENTATIONS AND SUMMARIES OF BREAKOUT SESSIONS

Following Dr. Pearson's presentation, workshop participants again broke up into four groups to discuss the issues of affirmative action backlash, lawsuits, globalization, and mentoring. Experts in each subject area introduced the topic and participated in the discussions. Each attendee participated in two sessions. The discussions were then summarized and presented to the entire group by one of the participants. The presentations and summaries of the discussions follow.

Affirmative Action Backlash

TYRONE D. TABORN
Chief Executive Officer
Career Communications Group

I like to think of myself as a journalist who somehow lost his way and became a manager. As CEO of Career Communications Group, which sponsors the Black Engineer of the Year Award, the Women of Color Award Conferences, Black Family Technology Awareness Week and La Familia Technology Week, and publishes *Black Engineer* and *Hispanic Engineer,* my subject today is affirmative action, specifically affirmative action backlash. After thinking long and hard about how to approach the issue of backlash, I decided to start with a question. What backlash are we talking about? Webster's defines backlash as "a sudden, violent, backward movement or reaction; a strong adverse reaction to a recent political or social development." But, there is nothing subtle or recent about trying to derail society's attempts to right the legacy of discrimination and racism in the United States.

Affirmative action has been subject to attack for its entire history. Political opponents have repeatedly tried to overturn it during the past 15 years. Now we have a secretary of labor and an attorney general who oppose affirmative action. So in fact, what we call a backlash against affirmative action has been around since the 1970s. The cover story of an issue of *U.S. News and World Report* in March 1976 was "Reverse Discrimination: Has It Gone Too Far?" Almost since the beginning, opponents of affirmative action have cloaked their motivations by calling the program reverse discrimination against white men. And that's been very effective with the public.

A lot of polls have been taken, but I'd like to mention one, a recent Gallup poll showing that 33 percent of white Americans favor a decrease in affirmative action. Only 25 percent of white Americans considered racism in the workplace to be a big problem. Among African-Americans, however, 56 percent felt that affirmative action should be increased, and 46 percent felt that racism in the workplace was a big problem. But what's behind these numbers continues to alarm me. Housing and educational patterns continue to show large-scale separation of the races, which leaves most images of African-Americans and Hispanics to be formed by the mass media. As a journalist, I've seen the impact of mass media on how we shape policy. Imagine that 56 percent of whites believe that blacks are less intelligent than whites and that 62 percent believe that blacks are less hard working. By itself, perhaps, this wouldn't be a big issue, but when it translates into hiring policies and teaching, the problems become serious. That's why those numbers are important.

Supporters of affirmative action seem to be losing the public relations battle right now. After a generation of progress, it appears that America's commitment to equal opportunity, not only for blacks but also for other minority groups and women, is at a crossroads. A movement dedicated to ending affirmative action, with a stream of electoral wins behind it, has led to significant rollbacks in equal opportunity programs. Somehow, opponents of affirmative action have perpetuated the idea that discrimination is a thing of the past and that affirmative action gives benefits to less qualified minorities, thereby denying opportunities to white men. The debate is often presented in the mass media as less qualified black people taking away jobs, college seats, and financial aid from more qualified whites.

Let's just take a moment to focus on education. Although minority groups are the fastest growing segment of the population, they make up only 12.4 percent of the entering class of doctors. Three medical schools, Howard, Meharry, and Morehouse, all historically black colleges and universities that play a major role in training black medical students, account for 15 percent of minority students entering medical school. Of the 123 other medical schools, six account for another 15 percent. So nine schools account for the enrollment of 30 percent of all minority medical students. Many of you might recall the Bakke case, challenging that 16 out of 100 places were held aside for qualified minority students by the medical school at the University of California Davis. The court ruled that, in the absence of past discrimination, quotas were illegal but that minority status could be used as a factor in an applicant's favor. The reality is that whites still receive the majority of top-level jobs, government contracts, and slots at leading colleges and professional schools. The facts show that affirmative action has not cheated white men out of jobs, wealth, and business opportunities. I don't recall a class action suit involving white men, but I'm aware of many settlements on behalf of minorities and women: Coca Cola, around $192 million; Texaco, $174 million; Denny's, $54 million; Mitsubishi, $34 million.

A report by the U.S. Department of Labor during the Clinton administration found that, of 3,000 discrimination opinions in federal district courts, reverse bias was an issue in fewer than 100. And even in those cases, the courts found that most of them involved a white applicant who was less qualified than the minority or woman applicant. So how is it that so many people have such mistaken perceptions of affirmative action? In some cases, we can attribute it to race-baiters who have convinced voters that affirmative action is not in the nation's best interest. But that only answers part of the question. Another factor may be the lack of public education. More than 60 percent of us look to television as our only source of information. We don't read enough, we don't question enough, and we're easily manipulated by politics of fear. So when it comes to a program for dealing with affirmative action and the workplace, education and debunking myths through the media must be major components.

Success will be critical for thousands of corporations in the coming years for several reasons. For one thing, litigation has increased as a result of recent civil rights legislation that gives more rights to class action suits as well as the ability to recoup recovery and litigation costs. That's why we see major litigators moving into discrimination law, and we're likely to see a lot more of that. For another thing, as many companies begin to lay off employees, minority populations will be adversely affected. In one of the best reports of its kind, back in 1992, a piece in the *Wall Street Journal* showed that during the last round of major layoffs, African-Americans were the group hardest hit. What is significant about this piece is that it showed that the African-American community never made up those jobs. Last hired, first fired. The Bureau of Labor Statistics shows that we seem to be heading to a repeat of those years, but the numbers have changed for the worse. I'll use March 2001 as a start, because during that month, layoffs increased, and unemployment of African-Americans increased to 8.6 percent, which was double the nationwide rate. Latino unemployment remained constant at 6.3 percent, which was still considerably higher than the 3.7 percent for whites. So corporations will be under great scrutiny to be just in their policies.

I recently spoke at The Alliance, which is the association of black employees of AT&T and NCR, where large numbers of women and minorities have been fired, while in those same departments, majority people were not fired. I can tell you that AT&T and NCR will be in court and litigation for quite some time because the stakeholders involved will be speaking out very loudly. Affirmative action programs may come under even greater attack and backlash as companies deal with their stakeholders. To help frame this discussion, I'll describe several issues that employers should consider.

First, affirmative action shouldn't be used as an excuse to explain away a lack of achievement and advancement. I remember my first year at Cornell. I came from the West Coast, from California, and I knew not a soul. I wound up at Telluride House, which was a scholarship house, where I believe I was the only minority person in over two decades. I was told by one of my housemates that "minority kids slow us down here at Cornell." In fact, one went further and said that minorities had taken his best friend's spot at Cornell. I was 18, and I don't think I knew much better. I wish I had known then what I know now. But, perhaps, my lack of knowledge helped put me on the path I have followed to find justice in this society. Only a week later, a young Jewish man befriended me there. It was clear from the beginning that he was not intellectually gifted, and he said, "Tyrone, I'm surprised I'm here at Cornell. But my mother and father went here. They're both doctors, and they gave Cornell a lot of money my senior year." So he wound up at Cornell. Like universities, at the executive level of most corporations, the selection criteria are often partly subjective and usually have little to do with GPAs, SAT scores, or performance ratings. Other considerations have much greater impact, such as having a high-level supporter, a coach, or a mentor who will speak on your behalf at the right time; being able to communicate and make presentations effectively; and making executives feel comfortable around you. In other words, understanding the corporate culture is key. I wonder how many complaints of reverse discrimination are simply the result of good old competition.

Everyone in the organization should understand that achieving diversity is in the interest of the company's business goals. Employees are just some of the many stakeholders in a corporation. Consumers, shareholders, and fund managers all have interests in the way large employers conduct themselves.

Third, companies must keep up proactive diversity actions. While a lot has changed in the way businesses operate, one thing remains the same—the need to have access to the best and brightest workforce. Companies can avoid tokenism by enlarging their workforces, which means traditional recruiting strategies will have to be augmented with programs that strongly emphasize work/life issues and emphasize the company's reputation as a good corporate citizen. To me this seems obvious. Companies ask me to talk to them all the time. The first thing I ask to see is their college recruiting schedule. Usually I say, "You're not going any place where there are Hispanics or African-Americans. How in the world do you even bring them in to be considered?" I always say that in business schools we don't teach Common Sense 101. Because people are usually comfortable with the places they come from, corporations tend to recruit from the same campuses where the top executives graduated. We've got to change that traditional thinking.

Finally, everyone in the organization should be involved in recruitment and diversity efforts, including the most senior executives. Put them on the campus at a historically black college or university. Involve them in outreach programs, in conferences with women and other minorities. Key executives should be part of both internal and external business resource groups. Companies have different names for these groups, such as women employees' groups or Hispanic employees' groups. All of these are business resource groups and should be looked at as resource groups.

In the end, perhaps the historical patterns of exclusion are so deeply ingrained in our society that this fight may never end. And perhaps that's good, because it means we will never forget. There are many bright and creative people who can and want to continue to make a difference. I was encouraged by a study from Ernst and Young showing that two-thirds of college students believe it is important to work for an organization that values diversity. Although that is promising, results will not be determined by words. They will be determined by actions. And those actions will necessarily generate reactions. I hope we will be successful someday in mitigating the reactions.

Summary of Discussion

Susan Staffin Metz
Women in Engineering Programs & Advocates Network and
Lore-El Center for Women in Engineering and Science,
Stevens Institute of Technology

We had a very active discussion about affirmative action and how to deal with backlash. We addressed three questions: What can employers do to avoid affirmative action backlash? How can employers respond to affirmative action backlash? How should individuals respond to affirmative action backlash? We focused on industry, specifically in the current political arena, in which we have a president, a secretary of education, and an attorney general who are essentially against affirmative action. Although we addressed three questions, we responded to two of the questions–how to avoid or how to respond to affirmative action backlash–in very similar ways. The consensus was that companies must be proactive, educate, communicate, involve all employees, and accentuate the positive.

Since its inception, affirmative action has been viewed as a quota system, and, according to Tyrone Taborn, supporters of affirmative action are losing the public relations battle. The issue is further complicated by the laws in California, Washington, and Michigan, which are completely inconsistent with federal laws. We also discussed semantics. Affirmative action has had such a negative connotation from the beginning that some people now use the term affirmative development. We need to put the idea of preferential treatment in the garbage and start anew. Also, sometimes the definition of the term *diversity* can be confusing. Diversity is an umbrella, and affirmative action is one of the programs under it.

In terms of education, a company must make sure everyone understands that diversity is a business goal, because diversity is good business. You have to make the business case, and you have to focus on qualifications. When you show pictures of people who have been hired or promoted, don't just show pictures. Include the titles and give some background on the people and their qualifications. Faces are not enough. Women and minorities shouldn't only appear in diversity publications. They should also appear in recruitment pieces, sales pieces, promotional pieces, and all advertising.

Recognition and awards programs have worked very well, but the criteria for hiring and promotion and the qualifications for awards should be known.

The consensus of the group was that excellence solves a lot of problems. On the corporate policy level, companies use newsletters, magazines, and e-mail within the company to make sure everyone gets the message.

We had an interesting discussion about how individuals should respond to backlash. Personalization of the issue works well at one level. People gave examples of managers, or their daughters or wives being the targets of discrimination, and all of a sudden, the issue became real. For some reason, that makes a big difference. The issue can't necessarily be addressed on the corporate level, but corporations can support external programs through professional societies and through mentoring programs, as well as inside the corporation. People have to let each other know how they can get this message across. We need to address backlash at all levels.

A Case Study of the Texaco Lawsuit

THOMAS S. WILLIAMSON, JR.
Chair, Texaco Task Force

I have been asked to address the topic of dealing with lawsuits. First, I will comment briefly on why employment litigation has become a growth industry in recent years. Then I will turn to my major subject, a case study of how a major oil company, Texaco,[1] responded to a lawsuit that generated national headlines in late 1996 by dramatically revamping its approach to managing diversity. By focusing on Texaco's experience, I do not mean to suggest that one size fits all in managing diversity, but the comprehensive nature of Texaco's response to the crisis the company faced illustrates a wide range of challenges and solutions that could be applicable to a variety of corporate employers, including engineering firms.

It is widely recognized and well documented in the legal community that there has been an upsurge in employment litigation (i.e., various types of employment discrimination lawsuits) in the past several years. In some ways, that trend seems counterintuitive. After all, the principal civil rights laws prohibiting racial and sexual discrimination have been on the books for more than 35 years, and a whole generation of CEOs and managers who had grown up either actively perpetrating or passively accepting racist and sexist traditions in the workplace as "business as usual" have either retired or simply died off. And, of course, virtually all businesses are now operating in a global economy where discriminatory practices seem especially outdated and counterproductive.

Notwithstanding our hopes that enlightenment in employment practices would have fully flowered by now, there are numerous reasons for the trend

---

[1]Texaco recently merged into Chevron, and the merged entity is now known as ChevronTexaco. However, I will refer to Texaco by its former name throughout these remarks.

toward increased litigation. One factor, of course, is that we have greatly expanded the number of protected groups since the mid-1960s. Women, who were added in the Civil Rights Act of 1964, have become increasingly active in the succeeding decades. The Age Discrimination in Employment Act, which was passed in 1967, defines the protected group as 40 years of age or older. Now that the majority of baby boomers are over 40, the pool of potential plaintiffs has swelled. And we cannot overlook Title I of the Americans with Disabilities Act (ADA), which was passed in 1990 and which protects against discrimination on the basis of disability.

Other factors have also contributed significantly to the rising trend in employment litigation. In 1991, Congress modified the federal civil rights laws relating to racial and sexual discrimination entitling plaintiffs to jury trials and compensatory damages for pain and suffering. Up to then, discrimination cases had been tried by judges, and the damages were largely limited to awards of back pay. Opening the field to larger damages greatly increased the incentives for plaintiffs' counsel, who are often compensated on a contingent basis determined as a percentage of the plaintiffs' dollar recovery. Finally, in a number of recent high-profile cases, plaintiffs have been awarded tens of millions of dollars or more. Cases involving Texaco and Coca Cola are among the most notorious cases in this category. In the Texaco case, the total cost of the settlement was about $176 million; in the Coca Cola case, the price tag for the settlement was even higher, $192.5 million. Those numbers sent potent business-opportunity signals to plaintiffs' lawyers who handle employment discrimination matters on a contingent basis, both at the class level and the individual level.

This trend in employment litigation suggests that the reasons for the upsurge are not likely to diminish or disappear in the near future. Therefore, employers must recognize that dealing with lawsuits that can have a material, adverse impact on their businesses is a long-term threat, not merely a temporary blip on the litigation radar screen.

Now, let's turn to the Texaco case. My knowledge of Texaco's experience is based entirely on my experience as a member and chair of the Texaco Task Force on Equality and Fairness during the past four years. Neither I nor my firm was involved in representing Texaco as defense counsel during the course of the litigation that resulted in the task force. However, as a participant in the task force, I have had extensive opportunities to learn about the circumstances that led to the litigation and the strategy Texaco adopted to turn the situation around. As a preliminary note, I want to make it clear that I am not participating in this conference in my official capacity as the chair of the task force. My remarks should be understood as my personal views.

Let me set the stage by describing how the Texaco litigation became a story that made national headlines. The lawsuit was a class action, race discrimination suit brought on behalf of salaried black employees of Texaco. The named plaintiffs were employees in the Texaco headquarters in Westchester County and in

New York City, who were involved in financial management matters for Texaco (investing pension funds). The media explosion occurred when an older, white male employee surreptitiously taped conversations of certain other white male executives who were responding to discovery requests from the plaintiffs. That employee turned the tapes over to plaintiffs' counsel, and they soon found their way to the *New York Times*, which published a story claiming that the taped executives had made racially derogatory statements about the plaintiffs, including use of the N-word. What had been poking along as a conventional, not particularly noteworthy discrimination case suddenly became a national *cause celebre*; and Texaco instantly became the poster boy for corporate racism. The value of Texaco's stock fell precipitously, and employee morale was decimated company wide.

Within a few days after the news broke, Texaco's management was facing a critical choice: (1) to defend the lawsuit and try to rebut the discrimination charges through the litigation process or (2) to settle the lawsuit promptly and use the litigation crisis as an opportunity to transform Texaco's corporate culture to foster and promote diversity. Of course, another option would have been for Texaco simply to settle by paying substantial damages to plaintiffs, thereby "putting the problem behind them," but Texaco agreed to pay substantial damages *and* assume responsibility for implementing comprehensive programmatic relief in the future. As I will explain, Texaco's commitment to prospective programmatic relief signaled that the company was not thinking simply in remedial terms, but was also thinking in terms of making fundamental changes in its corporate culture.

Perhaps the most dramatic evidence of Texaco's commitment to transforming its corporate culture was its acceptance, as part of the Settlement Agreement, of an independent task force of outsiders who would have the authority to design and oversee the implementation of the programmatic relief in the Settlement Agreement over a five-year period. In addition, the task force was given the authority to determine human resources policy for Texaco relating to fairness and diversity issues covered by the Settlement Agreement. This was a bold break with traditional principles of corporate governance, which abhor the notion of any control over management other than through the Board of Directors and limited types of shareholder initiatives. The new task force was to have a total of seven members, three chosen by plaintiffs, three chosen by Texaco, and a chairman who was jointly selected.

The Texaco case was settled in late 1996, and members of the task force were formally sworn in in June 1997 as special masters of the federal district court. In the interim period of about six months, between December 1996 and June 1997, Texaco publicly committed the company to undertaking a wide-ranging reform program that partly tracked the provisions of the Settlement Agreement, but also included ambitious efforts to promote diversity and hold executives accountable for successful performance on the diversity front. Notably, at the outset, Texaco's leadership made it clear that the reforms would not be targeted solely at salaried African-Americans (i.e., the class-action members) but would be intended to address the rights and needs of *all* employees, including women, white men, and all minority group members employed by Texaco.

By moving ahead with the design and implementation of a comprehensive program of diversity and fairness initiatives, Texaco preempted the possibility, allowed under the Settlement Agreement, that the task force would determine and design human resources policy for Texaco. Instead, the task force was able to carry out its mission and fulfill its responsibilities by acting primarily in an oversight capacity responding to Texaco's initiatives. This role also appealed to the task force because Texaco was more likely to stay committed in the long term to change that was generated internally than to directives that were imposed externally.

At the first meeting of the task force, the Texaco officials present informed the task force that the company had received widespread criticism in the corporate community for abrogating its management duty and "selling out" to plaintiffs' counsel. Notwithstanding the chorus of criticism, the Texaco officials announced that they believed the task force provided the company with a uniquely independent resource that they would not have been able to hire to assist them in fulfilling their responsibilities under the Settlement Agreement and achieving Texaco's broader equal opportunity and affirmative action goals. In other words, Texaco did not approach the task force as a punishment for past transgressions or as a Trojan horse that had been rolled into their midst. In the ensuing months and years, the relationship between the task force and the company has been spirited and candid, but always collaborative and collegial.

I will give you a brief description of five or six features of the Texaco program that struck me as particularly effective and likely to be transferable to other organizations seeking to develop a corporate culture that promotes diversity.

Executive Leadership. The top leadership, specifically including the CEO, must believe that promoting diversity directly advances the business interests of the enterprise and must make the case repeatedly and forcefully to middle management and line staff. Social justice and compliance with legal obligations are laudable and appropriate motives for encouraging equal opportunity and diversity, but there is a growing awareness that a strong business case is the key to sustained corporate commitment.

Expanded Recruitment. Expanding recruitment to include institutions where talented minorities and women are pursuing relevant technical degree programs is one of the most obvious elements of an effective diversity strategy. Texaco had traditionally drawn its engineering staff from Texas A&M and Penn State. Beginning in 1997, Texaco identified schools all over the country where the training of minorities and women in technological fields and business was an institutional priority. As a result, Texaco designated 40 "core" educational institutions for recruitment, including historically black colleges and universities and members of the Hispanic Association of Colleges and Universities.

Competency-Based Hiring and Promotion. Competency-based hiring and promotion procedures have substantially improved the fairness and quality of the job posting and promotion process at Texaco. Like most organizations, the people making decisions about hiring and promotion at Texaco tended to believe that people like themselves would be the best qualified. Texaco analyzed the various jobs in its workforce and systematically developed competencies for leadership and competencies for specific jobs. This process required about two years to complete. Subsequently, Texaco integrated these competencies into the job descriptions and interviewing protocols for assessing candidates for job openings and promotions.

Bonus Compensation. Another component of Texaco's transformative approach is the incorporation of diversity goals into the bonus compensation system for the 300 top executives of Texaco, whose bonuses are increased or decreased based on whether Texaco achieves its annual goals for increasing the representation of women and minorities in the ranks of the company's "professionals and managers." The assumption is that increased representation in those categories of employees will be likely to encourage an increase in the representation of women and minorities at all levels of the organization. The bonus-compensation formula communicates an ongoing institutional commitment to diversity and provides an accountability mechanism for diversity performance. Texaco has not always met its annual goals for increasing the representation of women and minorities, but the importance of remaining engaged in ongoing efforts to improve diversity is consistently reinforced through the bonus-compensation system.

Succession Planning. Texaco's commitment to diversity is also reflected in the company's approach to succession planning. Women and minorities regularly participate in the succession-planning process at all levels to ensure that all candidates who have demonstrated the potential for upward mobility into senior positions receive fair consideration. Also, in certain instances where there are gaps in the internal talent pool available to fill key positions, Texaco has retained search firms with the explicit understanding that they (1) must employ women and minorities as principals in their organizations and (2) must have a proven track record of identifying outstanding women and minority candidates.

Alternative Dispute Resolution. Finally, Texaco adopted an alternative dispute-resolution system that maximizes employees' sense of fairness and flexibility. This system, known as the Solutions Program, permits disgruntled or aggrieved employees to elect to have an outside mediator or arbitrator review their claims. If the employee elects the arbitration option, Texaco agrees to be bound by the arbitrator's decision that upholds the employee's position. The employee is allowed to pursue court action if he or she is disappointed with the arbitrator's ruling. The company has not found that its interests have been prejudiced by this asymmetry, and the program reassures employees that the company is committed to fairness in dealing with them.

As I indicated earlier, these are illustrative examples of Texaco's approach to diversity, but they do not present an exhaustive inventory of Texaco's fairness and diversity practices. I hope these examples are sufficient to give you some insight into what has been working at Texaco and what might be applicable to the engineering firms participating in this conference.

Summary of Discussion

Ray Mellado
Hispanic Engineer National Achievement Awards Corporation

The questions we were asked to address were: How can a company establish credibility in the aftermath of a lawsuit? How can corporate culture be changed? How can you get buy-in from the aggrieved people who were involved in the lawsuit?

At Texaco, some black employees had filed a lawsuit, which was moving forward when the *New York Times* got hold of a tape recording of a senior manager in an executive meeting allegedly using the N-word and making other disparaging remarks. This became national news, and the Texaco suit became the poster child for discrimination lawsuits across the country, even though it was an ordinary lawsuit much like suits against many other corporations. The Board of Directors decided to negotiate an agreement that would improve the company, change the corporate culture, and earn the buy-in of the litigants. In 1996, the suit was settled.

Six results came out of the settlement. The first was the change in corporate culture. The case was made that diversity was a good business decision for the company. Senior executives and senior managers of the company had to promote diversity. Middle managers had to have a process in place for managing diversity. All employees, minorities, nonminorities, and women, had to go through a diversity learning program.

The second result was the recruitment of minorities into both technical and nontechnical jobs, but especially technical jobs. In research, the majority of people hired at Texaco had come from two universities. As a result of the settlement, the recruiting pool was expanded to include students at 40 universities, including historically black colleges and universities and institutions serving Hispanics.

Three, the core competencies for hiring and promotion were publicized. The company set up a systematic matrix of core competencies for hiring and promotion. These are published on the web site and are open to the public and to employees. If an employee wants to move to the next level, he or she knows the required skills and criteria.

Fourth, bonuses and compensation for managers were tied to diversity. Five percent of the bonus structure was based on meeting diversity targets. If managers met the targets, they would get the bonus. If not, they would not get the bonus. This structure ensured some accountability and measurements of success.

The fifth result was succession planning. Every single meeting that deals with succession planning must be attended by at least one minority employee and one woman. This is unique, both for hiring and promotion.

The sixth result was an alternative approach to resolving disputes. The company brings in an independent counsel or independent person to mediate the dispute, and the company is bound by the mediator's decision. The employee, however, can go to court if he or she is dissatisfied with the outcome.

A key factor in the success of the program is the commitment of the Board of Directors. Having minorities and women on the board is important, but the entire Board of Directors must buy in to the program. Another key factor was for members of the protected classes and the litigants to become actively involved in succession planning and company planning. Now that Texaco has gone through this exercise for four years, they feel that what they are doing is good for all employees. The bottom line message is that smart planning and involving people from all backgrounds is good for everyone. I think the court did a good job in this case, as the results show. The real question is if the company would have gone through with this process if it hadn't been sued.

Diversity in the Global Marketplace

LISA NUNGESSER
Senior Vice President
Parsons Brinckerhoff Quade & Douglass

To prepare for this talk I did an Internet search on "global marketplace *and* diversity," and came up with 42,184 matches. I found that a wide variety of industries were publicizing their needs, making announcements, and selling courses; and the variety of firms was very interesting. There were a number of companies outside the United States and a number of global companies, ranging from transportation companies to Avon and Quaker Oats. Globalization is a very timely issue that is not unique to the engineering and design profession.

I'd like to define a few terms to clarify the framework of my remarks. The terms "global marketplace" and "globalization" are not in the dictionary yet. I would define the global marketplace as a borderless, multinational or multiregional place where goods and services are exchanged. An engineering design firm, which is in a knowledge industry, for example, sells professional services and knowledge. In *The Essential Drucker* (Harperbusiness, 2001), Peter Drucker, the noted consultant, says we cannot yet tell with certainty what the next society and the next economy will look like, because we are in the throes of a transition. But we do know that they will have a totally different "social complexion" from the current economy and society. Drucker's choice of words is illuminating— the new society, he says, "will be a knowledge society that has knowledge workers, and the largest, and by far most costly, element of business will be the workforce." That is the background for discussing the issues of globalization and diversity.

Why is globalization important in engineering? The easiest way to explain it is to look at the shifting market in construction. Back in 1975, the United States accounted for 50 percent of the revenues generated in the construction industry worldwide. In 1998, the United States still accounted for $900 billion in the revenue stream because of construction in the United States. But the worldwide market had grown to $3.2 trillion. Therefore, U.S. market share in construction, although very robust, dropped from 50 to 25 percent. I think that statistic is an easy way to get an understanding of the business imperative of going "global."

A man who works with me is on the board of a major construction firm, and part of his job is to deal with international engineering relationships. He raised three important points in our discussions of what a construction firm needs from us. In the design business, we like to see our energy turned into projects that improve the communities in which we work. Construction companies generally say that, at certain levels, all design firms are technically competent to do a specific project. The things the construction firms look for are quality, innovation, and responsiveness. And responsiveness includes both flexibility and speed.

How can companies, agencies, and industries deliver these to their clients? Of course, it's through people! The question is what kind of people we need and where we will get them. Bechtel is a long-time partner of Parsons Brinckerhoff and has been one of the biggest construction firms for many years. In 1980, a Bechtel executive gave a speech before the Associated Schools for Construction, in which he said, "We need people who are flexible, can effectively wear multiple hats, can work on large, complex projects, are people-oriented, and are prepared for business in the regulatory and litigious society that they are entering." I would add that it's also a diverse society. Notice that he didn't say they had to be good in science or good in math. In fact, he didn't even say they had to be good engineers. He just said they had to be flexible, wear multiple hats, be people-oriented, and be prepared for the business environment.

The underlying assumption of that speech was that engineering capabilities are developed through education, but these other skills are the ones you need to succeed. So, you can see that in a design industry we have both a tremendous opportunity and a business imperative to leverage our diversity to create an environment that will attract and retain top workers who can help us develop the leadership and innovation we need to deliver what our clients demand. Now I will turn to the experiences of our firm.

Parsons Brinckerhoff is one of the oldest continuously practicing engineering firms in the United States. We are a privately held company, with more than 9,000 employees working in 250 offices. We operate in 70 countries around the world, and more than 40 percent of our workforce resides outside the United States. That may surprise some people who work with us, but our historic base has actually been international. One of our founders started work on both the New York subway and the Chinese railroads. We have a long history of working beyond U.S. borders.

Why should you care what Parsons Brinckerhoff is doing? With some modesty, I'll tell you that we were pleasantly surprised that *Equal Opportunity* magazine, which is a career magazine for minority graduates, recently picked

Parsons Brinckerhoff as the fifteenth best employer for minorities in the United States. This was not just in our peer group of engineering design firms. We were listed with companies like Microsoft and Dell and Intel, and we were ranked higher than McDonald's, Ford, and Yahoo. Of all of those companies, we were one of the best for minorities. In our own peer group, just this month *Civil Engineering* magazine rated Parsons Brinckerhoff the eighth best place for civil engineers to work. The competitors that did well were smaller companies, including CH2M HILL, which is a very good employer.

I want to highlight a few of the programs at Parsons Brinckerhoff that are especially important for global diversity. In 1997, our CEO and human resources director decided that so many factors affect our workforce beyond the traditional human-resources factors that we should establish a committee that would answer directly to the CEO on diversity issues. We now have four focus groups to help with employment issues, including development issues, recruiting and retention related to women, African-American, Asian-American, and Hispanic-American employees. The women's group is the oldest, and we sometimes joke that PB stands for "pushy broads." This group has tackled one issue related to global diversity—that the company had no women of any stature outside of the United States. We did a survey of managers outside the United States to determine their attitudes and found that their attitudes toward women were positive. Managers who had worked with a woman overseas were receptive to the idea of international assignments for women. We then surveyed our women who had worked in our offices abroad and used the list of women who thought things had gone well for future assignments. The company subsequently adopted diversity policies for our global presence that reflect our climate and our managers' expectations.

We noticed that our turnover rate for young professionals was a lot higher than we wanted, so our chairman went out to lunch with a group of young professionals to find out what was on their minds. He learned that they felt they didn't have enough exposure, didn't have enough leadership opportunities, and didn't have enough say in how the company was run. To address their concerns, we created a professional growth network that includes employees in the United States, Asia, and Europe. The participants, who have less than 10 years of experience, have taken on improving college relations, sponsored competitions, provided mentors, and welcomed new people and shown them how to get things done. Our turnover rate has dropped, and the program has given us access to the next generation of leaders.

We have also developed practice area networks. We now have about 50 of them related to the disciplines of the people we hire. There might be one mechanical engineer in Tampa and one in Seattle, for example, both in big offices, each one the only representative of that technical specialty. But they do not feel isolated because they are part of a network of mechanical engineers around the world. These networks also are great resources for employees who need advice. We have created networks in 50 different disciplines, and, all of a sudden, we had created a knowledge-based company. For example, let's say a client asks us if we have any experience using license plate matching for origin destination studies with a certain camera technology. We can send a request out on the appropriate company networks, and we may get 15 e-mail messages back describing experiences in various places. We can then gather the information and pass it on to the client.

I have highlighted only a few of our many programs. The global marketplace is already at the door, bringing tremendous opportunities for all of us in engineering fields.

Summary of Discussion

Gary Downey
Virginia Tech

Our topic, globalization, is not easy to grapple with. We were fortunate to have an excellent presentation by Lisa Nungesser from Parsons Brinckerhoff, which is out front in thinking through the question of global diversity. We had a wide ranging discussion with some significant disagreements, especially in the first session, and a combination, I think a healthy combination, of optimism and pessimism. Sometimes the optimism was stated more explicitly than the pessimism, but both were clearly present.

First, we pieced together some questions from the various comments. What sort of phenomenon is globalization? What new demands do we face because of globalization? What are the key characteristics of workers in a global marketplace? We spent most of our time in both sessions talking about the interesting challenges facing local divisions or local companies and branch offices located offshore or outside the United States. Finally, what best practices exist in industry and in education?

Globalization involves doing business in a borderless market, a world in which workers are valued for their knowledge, which calls attention to their education and training. Peter Drucker describes the emerging global workplace as being characterized by a new "social complexion." The pun regarding race and gender is probably intended. Globalization involves a larger dimension of diversity, and one person suggested that we think about it in terms of "inclusiveness." Does diversity mean different things in different places? Clearly it does. To the extent that it does and to the extent that there are new dimensions under the label of diversity, we have to be careful to keep the problem of underrepresentation and inclusiveness in mind and not let them be overlooked because of important cultural differences. What does globalization mean for the poor? Is globalization an exciting opportunity? That is one area where we had significant disagreements, but we didn't pursue the subject very thoroughly. Finally, one person suggested that globalization is absolutely necessary for building a sustainable world.

What sorts of new demands are created by globalization? In the construction market, the global market share for U.S. companies has dropped from 50 percent to 25 percent. So our organizations must focus intensely on quality and responsiveness. There are significant variations by industry, which is certainly an important area to study. Globalization also raises concerns and creates dangers. Is there a loss of jobs or a potential loss of jobs for U.S. engineers? What are the implications? What about domestic education? If we are competing globally for a fixed labor pool, isn't this another reason for expanding the domestic pool of engineers? Finally, it is important, especially in the wake of September 11, that we understand global diversity inside the United States. People from many countries died on September 11, but in many places non-U.S. born populations in this country are relatively invisible.

Worker characteristics in an environment of globalization must include not only technical capabilities, but also other sorts of skills. Workers must be people-oriented, able to wear multiple hats, and be well versed in the business, as well as the social and litigious nature of today's society.

We spent most of our time discussing the issues facing the local division of a company operating outside of the United States. Each company operating outside the United States must address the question: Am I a visitor, or am I here for the long haul? If I am here for the long haul, I must make a serious commitment to accommodating myself to the environment in which I am working. But the issues vary from country to country, and companies must tailor their programs appropriately. Companies need a top-down commitment to inclusiveness, whatever that means in the local context. Even the word *diversity* can get us into trouble. The word means different things in different places, and we need to be more sensitive to that. This is an emergent issue that was identified in a recent Catalyst report, *Passport to Opportunity: U.S. Women in Global Business* (2000).

The second group spent some time discussing the problems of two-career families. What happens when a spouse, whether female or male, would also like to work in another country? We also spent a lot of time talking about what I would characterize as "somebody has to understand the Russians." Bob Spitzer talked about Boeing's experience in Russia and described the qualities of the Russian engineers, who are quite good and sometimes even embarrassed the Americans. It is important that we understand these people.

Over the course of two sessions, a number of examples of cultural differences were raised. A woman leader is perceived differently in Taiwan and Brazil and in Japan. We heard a story about an e-mail joke in South Africa that was considered very funny in South Africa but would be considered sexual harassment in the United States. Malaysia has hiring quotas, and you had better follow them. The population of Turks in Germany is increasing. Should global diversity managers be knowledgeable about demographic trends and try to build their companies in a way that anticipates demographic changes? Workers in London tell us to stop sending Americans to be their bosses. That is a diversity issue for Duke Energy. If we sell airplanes to Spain, we should hire their workers as well; we had better locate some kind of a production and research outfit in the country. As we develop linkages, many different types of questions will arise that will involve our understanding perspectives that are different from ours. This was a common theme in our discussions.

Finally, we must be careful about imposing U.S. expectations on people in other countries. The cell phone and beeper are anathema to engineers in some cultures, and the idea of being accessible 24 hours a day is repugnant to some people. There was a brief discussion about the current European slow movement, which is characterized locally as resistance to Americanization. Many Europeans understand globalization as Americanization.

I'll give you a quick summary of a few of the best practices we discussed. First, it is important for companies to conduct research across divisions, to constantly monitor and understand the organization. Second, it is important to develop a global diversity policy. Parsons Brinckerhoff, for example, developed a professional growth network by country, not continent. It is possible now to build practice-area networks, in which all of the mechanical engineers knowledgeable in such and such an area can form a virtual group and mentor one another across national boundaries.

A big issue for any company that does business in more than one country is certification. The requirements for engineers and licensed engineers varies from country to country. Europe, in particular, is struggling to develop a concept of the European *ingenieur*. I am purposely using the French word for engineer, but the big challenge was how to license British engineers, which the French and Germans opposed. Certification is a difficult issue. A global diversity policy should be incorporated into every division of the company throughout the world.

We spent a fair amount of time talking about language training. This summer I took a group of women undergraduate engineering students to Paris for two weeks. We visited Renault, where a woman engineer, who had come originally from Purdue by way of Germany, and is now a senior manager, argued for the importance of language training. Until she developed language skills, she said, she had trouble finding work in Europe. Joan Straumanis from the U.S. Department of Education described the University of Rhode Island double-degree program in engineering and German, specifically for German-speaking engineers. It is important to preserve heritage languages so these are not lost. We need to take advantage of people's capabilities, including their knowledge of languages.

This raised the general issue of helping engineers to be open to other cultures, which led to a discussion of engineering and the humanities and a course called Engineering Cultures that is now available, or soon will be available, as a continuing education tool for engineers interested in learning about Japan, Europe, and Russia; down the road other modules will be available on Korea, China and Taiwan, India, Mexico, and Brazil. The course is designed specifically for continuing education purposes, and the goal is explicitly to help engineers understand other engineers raised in other national cultures. The course will also be available as modules or as a one-semester course for students at the undergraduate level. Lastly, *Great Leaders See the Future First* (Dearborn Trade, 2000) by Carolyn Corbin, was recommended to us. Corbin includes information on demographic trends in her discussion of global diversity.

Mentoring

JANET M. GRAHAM AND SARAH ANN KERR[1]
E.I. du Pont de Nemours and Company

As an introduction to the topic of mentoring, Janet Graham and Sarah Kerr of E.I. du Pont de Nemours and Company presented an overview of DuPont's mentoring program. They described the four guiding principles of the DuPont mentoring program, which are that it is:

- voluntary
- open to all employees
- protégé driven
- management supported with encouragement and administrative resources

The program is protégé driven in the sense that individuals who want to participate as protégés (or mentees) are responsible for identifying their preferred mentors and approaching them with their requests.

Based on their experience in DuPont's mentoring program, the participants have learned that successful mentoring relationships require that both parties be interested in learning from each other and that both believe the experience of mentoring is valuable, to the individuals involved and to the company. Discussions between mentors and protégés must be confidential, and the parties must treat each other with trust and respect. Each mentor and protégé pair should agree on the goals of the relationship in advance and keep their commitments to each other.

---

[1] Ms. Graham and Ms. Kerr's remarks were summarized by NAE staff and approved by the authors.

A successful mentor should be easy to approach, a good listener, committed to his or her personal development and to the development of others, a provider of honest, constructive feedback, and should respect different perspectives. Because mentoring provides mentors an opportunity for reflecting on their career experiences, learning is often reciprocal—the mentor learns from the protégé and vice versa. A senior employee who acts as a mentor can improve his or her listening and interpersonal skills and become more appreciative of cultural diversity. Mentors often feel pride in contributing to the development of junior colleagues.

A successful protégé must be willing to learn from the experience of others and must be able to accept feedback. A mentoring relationship provides an opportunity for employees to receive constructive, unbiased feedback outside of the regular performance review process. Protégés have an opportunity to gain a better understanding of the corporate hierarchy and to learn the unwritten rules of the organization. They often gain confidence, competence, and credibility through participating in a mentoring relationship. The interaction between mentors and protégés can also bring new perspectives to the organization and challenge organizational thinking.

Perceived benefits to the organization of a successful mentoring program include employee development and an effective transfer of institutional knowledge. Relationships between employees with different levels of experience and expertise often generate new ideas and encourage cultural exchange. Employee satisfaction can lead to higher retention rates, increased productivity, and can ultimately give the company a competitive advantage.

DuPont identified eleven factors necessary for a successful mentoring program. During the start-up phase, the planners should collect information on mentoring to take advantage of the experience of others. The company should put together a design team to identify the needs of employees and set goals for the program. A coordinator should be appointed to oversee the program and keep it going and to enlist the support of management. Before implementing a full-scale program, the company should initiate a pilot program in a single department or small group of employees. The program should be marketed to potential mentors and protégés to ensure that everyone understands the mechanics and benefits of the program. Pairings should be based on a matching process that allows both mentors and protégés to specify their preferences and expectations for a mentoring relationship. All participants should be trained in ways to make the mentoring experience productive. The supervisors of participants should be involved in the program and should be aware of the benefits. The company should have a mechanism for keeping track of the mentoring relationships, so that the results can be evaluated. Finally, succession planning should be part of the program.

Some of the pitfalls and challenges of mentoring programs identified by du Pont are: meeting too infrequently, lack of meaningful discussions, and failure to follow up on important issues. Mentoring relationships that cross gender, ethnic, or cultural lines can be especially challenging. In short, the mentor and protégé should be aware of potential problems and prepared to address them.

DuPont had little experience with measuring the success of their mentoring program quantitatively. Feedback and testimonials from participants, however, provided a qualitative assessment of the program, which was mostly positive. Some mentors commented that the program had increased their appreciation of differences and their trust in others. Mentors also felt that their participation had changed their perceptions and their understanding of others and of themselves. Protégés felt they had gained a better understanding of the organization, had expanded their networks, had increased their sense of empowerment and had raised their level of self-esteem.

Summary of Discussion

Suzanne Brainard
Center for Workforce Development
University of Washington

Our topic was mentoring programs. The three questions we addressed were: What are the key components of effective mentoring programs? Are there potential pitfalls in mentoring programs for diverse employees, and, if so, how should they be handled? Is there a "best" way to deal with the issues of gender and ethnicity in mentoring programs?

Most people felt that companies should specify their goals and objectives for mentoring programs. Is the objective to improve recruitment? to increase retention? or to promote the advancement of diverse groups? Mentoring programs would be implemented in different ways depending on their objectives. Another critical component was commitment from the leadership in the company or institution and the communication of that commitment to employees throughout the organization. DuPont's diversity programs were given as examples. Diversity and mentoring were discussed with employees and through external seminars for deans of colleges of engineering. The University of Washington's Curriculum for Training Mentors and Mentees in Science and Engineering, which is now used in 250 institutions across the country, provides a designed approach to implementing mentoring programs and a methodology for evaluating their effectiveness. Another critical component of effective mentoring programs was accountability or a method for measuring their effectiveness in terms of retention, recruitment, and advancement.

The second question addressed potential pitfalls in mentoring programs for diverse employees and how can they be avoided. Several members of the group noted that research has shown that gender and cross-racial mentoring can be effective when the participants are trained and educated in advance about sensitivities to diversity. Some of the literature concludes that cross-racial and cross-gender mentoring can never be effective. Take, for example, an African-American female mentee and a Caucasian male mentor; this situation could be fraught with complicated issues. The combination is not necessarily negative if both have been made aware of differences and expectations based on cultural experiences, as well as gender experiences. Another pitfall was "terminating a relationship." In both academic and corporate settings, mentees are often easily intimidated by senior mentors, regardless of their race or ethnicity. As a result, connecting with a senior mentor may be difficult and often delayed by the mentee. Sometimes, the "chemistry" is just not right. Mentors and mentees must be given the option of terminating the relationship for whatever reason and of being rematched with no fault attached to either. Mentoring programs that have built in "fault-free termination" are more effective.

Finally, the third question was if there is a "best" way to deal with cross-gender, cross-racial, and similar issues in mentoring programs. One of the most effective ways of matching mentors and mentees is to have them identify the desired characteristics of their partners in terms of race, gender, age, sexual orientation, and so on. Doing this at the beginning may avoid some pitfalls. Another good idea is to build a feedback loop into the mentoring program so that mentors and mentees can relay problems as they arise. Having a coordinator for the program is critical then, to ensure that problems are addressed up front. Finally, training in mentoring and educating participants to the sensitivities of people from diverse backgrounds and cultures is very important to ensuring that the relationships are beneficial to both mentors and mentees.

Morning Keynote Address

Implementing Change

NICHOLAS DONOFRIO
Senior Vice President, Technology and Manufacturing
IBM

My passion is for engineers and scientists, the people who generate real wealth in the world. I think we need more of them. The need for technical talent is clearly a critical issue in my industry, the information technology industry. Yes, we are going through a bit of an up-and-down in the economy and, yes, this will be a very difficult recruiting year for young women and young men on college campuses around the world. But we will get over that.

What's scary, though, is the long-term trend in the information technology industry toward huge shortages of engineers and scientists. Predictions are that there will be a shortage of perhaps two million, globally, within five years. If they are right, the debate over H-1B visas will be silenced because there will be no one to come here to solve our problems. Clearly, the solution to our problems is to generate more engineers and more scientists, to encourage people to focus on careers that can literally generate more real wealth in the world.

My bona fides are pretty straightforward—34 years with IBM. I am an engineer, an electrical engineer. I did honest work for a living. For almost half of my career, I did real things, made real things, designed real things that actually worked and generated wealth for IBM. For the last half of my career, you will have to excuse me, I have been an executive involved completely in management.

I have been a member of the board of the National Action Council for Minorities in Engineering (NACME) for 10 years. Bill Friend, a member of the NAE Committee on Diversity in the Engineering Workforce, is a personal friend of mine, and he is the reason I am chairman of the board of NACME. I am just coming to the end of my four-year stint, so I will soon be passing the baton on to someone else. I have also been involved with National Engineers Week (NEW) for perhaps the last seven years, as IBM's executive contact. We chaired NEW this past year on its fiftieth anniversary. All of those things put together mean something to me and are part of what we at IBM consider best practices. I recommend them to you for making a difference in engineering, science, and diversity.

Diversity to me means gender diversity, racial diversity, and ethnic diversity. I am just as passionate about women in technology as I am about underrepresented minorities in technology. You know what the numbers are so I am not going to bore you by spouting data. If you don't know what they are, NACME has a wonderful publication called the *NACME Journal*, which has all of the latest data. And the data are incredibly disturbing. Over the past 20 years, all of the progress we have made was made in the first 10 years; we have made virtually no progress in the past 10 years. That is a horrible indictment of me and my involvement, but it is unfortunately true. When you look at enrollments and graduation rates, we have gone nowhere. The numbers have flattened out.

It is time for us to think differently about what we are doing. We now know we cannot expect to make continuous progress. So the best thing to do, based on my industry experience, is to stop doing what we have been doing and try to be creative and think "out of the box." We have to start coloring outside the lines and come up with some type of breakthrough thinking.

You know, I find it very stimulating to come here. Number one, I love the National Academies Building. Number two, as most of you probably know, this is the birthplace of NACME; it was born just outside of the Great Hall in this building about 27 years ago, when the National Academy of Engineering recognized that diversity was going to be a huge problem and that we had to do something about it. So, they met and created NACME, which started out with an incredible agenda and incredible objectives. Unfortunately, I am very saddened to tell you, we missed them all. We missed them all. Back then, they thought that by this point in time we would be a $25 million organization that would generate at least 20,000 underrepresented minority engineers in this country. Well, we are not. We are barely half that. We are a $10 million organization, and we have supported 10,000 underrepresented engineers. Not a total failure, of course. We wouldn't have made the progress we have made without NACME, but we are not getting the job done.

I don't know the other organizations as intimately as I know NACME. I know GEM (the National Consortium for Graduate Degrees for Minorities in Engineering and Science, Inc.) quite well because IBM is deeply involved with GEM. I know the Hispanic Engineer National Achievement Awards Conference, and I know some others, but not as intimately as I know NACME, which is in my blood after 10 years. I also know that NACME is one of the most organizationally competent organizations in this community. NACME's present leader, John Slaughter, is an incredible force for change as he has been for his entire life—when he was head of the National Science Foundation, chancellor of the

University of Maryland at College Park, and when he was president of Occidental College, where I met him.

Given our lack of progress, the growing need for change, and perhaps the right leadership, even in these tough times, we may be at the right point in time to make significant changes. I believe that if a company has the right vision, now is the right time to make substantive changes. Look at the demographics. This is the perfect time. Very soon, there is going to be an abundance of brilliant students—women and underrepresented minorities—who are going to be looking for something to do, and there are not going to be enough jobs for them. They are either going to go back to school, or, I hope, IBM will hire them all and make a substantial change in the mix in the company.

We are very proud of our diversity efforts. I am not going to spout off about them to you, but we don't come late to this arena. IBM has been fighting in this arena for 80 years, because that is the way we grew up under Mr. Watson, Sr. This was his idea. He wasn't confused about morals and ethics. He wasn't confused about business. He just thought diversity was the right, rational, sane thing to do. After all, we are the *International* Business Machines Company. We have always liked the idea of diversity of thought—we operate in 160 countries around the world. We have eight research facilities in eight different countries around the world to capitalize on diversity of thought. We appreciate the fact that women see problems a little differently than men. We appreciate the fact that underrepresented minorities bring a diversity of thought and creativity to thinking about problems and solving them. We have no trouble getting behind these types of efforts, but there are lots of problems.

I'll give you my perspective, based on the three hats I wear. I am chairman of the board of NACME, a senior executive with NEW, and a senior vice president of IBM. From a NACME perspective, we cannot allow ourselves to be continuously fractionalized and marginalized in our efforts to address diversity in engineering, but we do. Too many of us are asking the same people for the same things, over and over again. We can't afford to keep doing that, especially in bad times, because they will begin to throw darts at the board and say "yes" to this and "no" to that. Why? Because they are simply out of money. We can't keep dividing our efforts into finer and finer pieces. It doesn't make sense.

First let me talk from a NACME perspective. As some of you might know, I spoke at the closing banquet of NACME's "Forum 21" in Baltimore on Saturday night, and my charge to the organization and to John Slaughter was simple. We have to find a way to collaborate more, align ourselves more, perhaps even consolidate some of the myriad organizations that seem to be saying the same thing to corporate America, because corporate America is going to be an incredibly difficult place to get support.

Even in NACME, commitments from incredibly strong, capable, well intentioned, well meaning corporations are falling by the wayside because they just don't have the resources. That is going to be the reality for the foreseeable future. We have to deal with the issue of too many of us saying too many of the same things to too many of the same people. We have got to align our efforts in some way.

I think now is a good time for us to think about ways of getting our agendas moving in the same direction. John Slaughter understands this and is moving in this direction. If ever there was a right person, a collegial leader, who was willing to give and take, to give and get, it is John. So, I am hopeful. I pray that we can pull something off here. By the way, Bill Wulf, president of the NAE, sits on the board of directors of NACME. This shows that we are working together in some ways, but we need to do much more.

My second point of view is from NEW. I don't know if you are aware of the value of NEW, which is a great force for change in this country. I like it because it gives us an opportunity to get back into the school system and do something about math and science so we can influence the outcomes, the possibilities, the probabilities, the options for young women and young men from underrepresented minorities in the elementary school system, which is incredibly important for us. In grades 4, 5, 6, and 7, most children make career-altering decisions. They are deciding they hate math or they don't like science or they don't like the science teacher or they hate the math teacher and, therefore, hate math. These are career-altering decisions. These same young women and young men will then tell you they want to become doctors or astronauts or engineers.

They see no linkage, no cause and effect. They have no idea that what they do now will affect their futures and change their options. That is why I like NEW. It lets me, lets us, lets IBM, lets all of you who participate get back into the schools and show kids what can happen, how real wealth is generated, how exciting being an engineer can be from a math and science perspective. You can make a difference. You bring your liquid nitrogen, your polymers, your smoke and mirrors and you show young people that they can make a difference in the world. This is what engineering and science is all about.

Last year, IBM chaired the fiftieth anniversary celebration of NEW. NACME, NAE, and many of your companies have been partners with NEW. Many of you have chaired NEW activities. My only frustration is that NEW's presence in the schools only lasts for two or three weeks. We have to find a way to stay invested in our K through 12 system, especially in grades 4, 5, and 6.

Perhaps the proudest part of IBM's involvement with NEW is "Girls in Technology," an educational module that was rolled out across the country in 2001. As a result, tens of thousands of young women in grades 6, 7, and 8 had the opportunity to work in a technology-oriented company to see what engineering and science are all about. We developed the module three or four years ago at IBM and then passed it on to NEW.

The third perspective is from IBM, where I am responsible for IBM's technology, and for our worldwide population of technical professionals. IBM has

160,000 technical people out of a company of 320,000 worldwide. We are 160,000 strong—engineers and scientists in every discipline, including software, services, hardware, manufacturing, research, and development in the field and in the support structure. We have an incredible program to attract, retain, and develop people, to address their cares and concerns. Two programs that I am very proud of deal with diversity, in terms of ethnicity, race, and gender. One is the Women in Technology Council; the other is called the Multicultural People in Technology Council.

IBM created these two groups for technical people to give them a forum for expressing their ideas. The main thing we want to learn from all of our diversity councils is why they want to be at IBM. What does it take to make them happy here? Do they feel comfortable here? Are they at home here? Are they uncomfortable here? If so, how can we fix it? Women in Technology brings together women from 160 countries. The common bond among these women is that they feel their technical contributions are not being recognized. I didn't lead this initiative, but I started it and I sponsor it. Remember, senior executives have to sponsor these things.

I thought we could do the same for African-Americans, Hispanics, and so on, but then I realized we could end up with a lot of separate groups. A bright young woman in Raleigh came up with the idea of having a Multicultural People in Technology Council. The group includes people from all minorities—Asian-Pacific Islanders, African-Americans, Latinos, Latinas, Native Americans, and many others. Once the group got started, it began to create its own agenda, hold its own conferences, build its own networking structure, build its own mentoring network. All it took was a little bit of help from me, a little bit of money, a lot of nurturing, and a lot of publicity.

We created these two councils for our own preservation, not because the government said we had to have them. IBM needed them to maintain diversity of thought. Our technical people respect diversity of thought.

We are the most inventive company in the United States, and have been for the past eight years, and I guarantee you, we will be for the ninth year when the books close for this year. We have built an environment where people feel somebody is listening to their concerns. Now, we are not perfect by any stretch of the imagination, and I am still not happy with our numbers or with the mix of people. They still don't look right. There aren't enough women or underrepresented minorities getting some of the $3 million a year we give at our awards conference to especially gifted people. We also have a 300-person academy of technology. I don't like the mix there either—not enough women or underrepresented minorities. It is always dangerous, of course, to try to manipulate things. Technical people go nuts if they feel the heavy hand of management reaching into their pockets, taking their arms out, and making them vote for this or that. But we will make progress. I guarantee you that before I retire we will make substantial progress.

I always harken back to a famous book written by a famous IBMer, Fred Brooks, a founding inventor of System 360. He worked in Poughkeepsie, and he didn't spend a lot of time with IBM, but he did some incredibly bright work. When he left IBM, he went to work for the University of North Carolina at Chapel Hill as a computer scientist. I think he has just retired. In his book, *The Mythical Man Month* (Addison-Wesley, 1995), he simply says there is no silver bullet for software. There is no silver bullet, no magic potion that can fix things. Things only get fixed if you take the problem personally, and you want to make a difference. It is that simple. If you care enough about it to put your money where your mouth is, and you put somebody in a leadership position to do something about it, you can make a difference.

I hear from our chairman all the time asking why I can't fix things right away. But fixing things takes time and sustained effort. In addition, you simply have to take things personally. That is my story whether you like it or not, and I am sticking to it.

Questions and Answers

Gary Downey (Virginia Tech): As someone who works in a world of technical people, what is your take on engineering education in this country today?

Nick Donofrio: Actually, I am pretty hopeful, pretty optimistic. I should have also told you I am on the board of trustees of Rensselaer Polytechnic Institute (RPI), which is my alma mater. I got my master's degree from Syracuse University, but I never actually attended Syracuse. I worked full time and got my degree at night. So, I grew up the hard way. I was at RPI when it was a tough school, when the dean or the president looked at us and said, "Look to your right, look to your left; one of you won't graduate." And they meant it. We were lucky that half of us got out the door, but they delivered at RPI. I don't think there is anything wrong with engineering education in our college and university systems. I am incredibly impressed and awed by the brilliance of new graduates at IBM.

I will also tell you that they are happier and more well rounded than any students I have seen before. Perhaps one thing we can do a better job with is convincing engineers and scientists that part of their mission, part of their job is to explain complicated things simply and succinctly. I heard a wonderful lecture on quantum computing just the other day by a guy in the forefront of the field. He had simplified quantum computing into an incredibly understandable presentation that ignited me and the audience to want to learn more. In our lifetimes, we might actually switch to the quantum model. The speaker was an IBM Fellow, very qualified, with a huge portfolio of patents. He barely comes out of his office. But, lo and behold, he did a brilliant job of simplifying a very complex subject. I think it is critical that we be able to express ourselves.

We need to focus on that more because we tend to think the more difficult a problem is, the more we say about how complex it is, the more jargon we use the better we are doing. But that is not true. We have to simplify things, to get them down to understandable facts.

I am not saying everybody has to be a business person. I have already told you we only need a few business people to count the money in the counting room. We do need a lot of people who can make things and generate the money. Usually, we can teach people the business skills they need. I am not against MBAs, but I have been incredibly successful without one. We need more engineers and more scientists who can sell their ideas, communicate their ideas, talk about their ideas.

Today's engineering students are more gifted than ever. I would never be hired back into the company if I graduated with my 3.1 GPA from RPI this year. We typically don't even look at kids unless their GPAs are in the 3.4s or 3.5s. I would have had to do something incredibly important to get IBM's attention now. And I worry about that, too, by the way. We spend a lot of our time on campus talking to professors and department heads asking who the brightest people are. We know GPAs don't always correlate with success, but it is the only thing we can use to screen people unless we have qualitative input from deans or department chairs or the students have experience in co-ops or as interns.

But in general, I like the education system. But I want people to be a little more articulate—not necessarily charismatic, but articulate.

Jim Johnson (Howard University): You spoke of the need for organizations to address the need for more minorities and women in quantitative fields. But I also think there is a need for corporations to get together in a coalition to provide some building blocks to attack systemic problems in cities.

For example, in Washington, there are at least half a dozen corporations, but their lead line is here is my product, we want you to use it, rather than let's identify problems and see how we can pool our resources and make changes and then use our products as a way of demonstrating or a way for students to practice how to improve and move forward with the change. What do you think about a coalition of companies coming together to attack systemic problems and investing their resources in a way that puts their names up front?

Nick Donofrio: I would have to agree with you. IBM, under Mr. Gerstner's leadership, has shifted its philanthropic giving to education reform in K through 12. We built a national alliance to bring governors together because we believe education at that level is a state issue, not a U.S government issue. It may even be a city issue and not a state issue.

The alliance is a self-perpetuating group for reforming education. Mr. Gerstner believes in rigorous standardization of instruction and of certification, so he looks first to the administration, then the faculty, and then the students. This is a very old formula of discipline in structure that says we know what we are teaching them, and we know they are learning something. Let's invest our money in those parts of the system. Physical structures are just physical structures. Let's invest our capital in people, whether in the administration, the faculty, or the students. That is our K through 12 initiative in a nutshell.

Other companies have joined us, but we have not created the kind of alliance you are talking about. There is no real pooling of energy in the alliance other than conferences once or twice a year to share ideas. We run our own K through 12 program, G.E. runs its own, Ford and GM and DuPont and everybody else runs their own. There is no sense of togetherness. But it is not a bad idea, though, maybe something whose time has come.

But K through 12 education has clearly got to be the centerpiece of any reform. That is where all the problems are. We can't fix them at the college and university level because it is too late. They are what they are by then. Thank God for Howard graduates, but the fact is we are not getting enough good people. I would like to see more cooperation with colleges and universities to improve K through 12 education. The idea has fallen on deaf ears. Colleges and universities seem to be more interested in buildings. They offer us naming opportunities for only $5 million. I say we don't want a name on a building and we don't have five million bucks. We tell them if they don't get in line on our K through 12 initiative, they will get no money from us.

In the nine years Mr. Gerstner has been with us, we have not funded college and university activities. We participate in partnerships, and we run programs. In fact, I run the university relations program. We give grants for things that a university does for us, so, we do share in some university research projects. But I don't want to see a Nick Donofrio Gymnasium or a Lou Gerstner Memorial Auditorium

Ray Mellado (Hispanic Engineer National Achievement Awards Conference): I have two questions about information technology. A couple of weeks ago I heard Robert Card, the undersecretary of energy, and Irving Wladawsky-Berger of IBM talking about security at our nuclear and DOE facilities and national labs around the country. What do you think about information technology security in the short term and long term and how do you see it evolving?

Nick Donofrio: There are lots of things that can bring us to our knees in the information technology industry. People happen to be first on my list. We don't have enough talent, and we don't get the job done. I think that is clear to you.

Second is security and privacy, which I put together. The whole world is on line, and e-business thinking is real. The dot com mania is over, and we are going back to the real business of business. Electronic communications and commerce is the way business is going to be structured, both inside and outside. There are people here in Washington who are proposing to build their own

Internet with a different set of standards. I think that is nuts, to be honest with you. We don't need that. We need more secure systems and better administrators. We need better enforcement. I attended President Clinton's conference of industry leaders at the White House after e-Bay was attacked with a denial-of-service attack.

The single biggest problem with security systems worldwide is that nobody uses them. Our own experience suggests that one-third of the customers we service never change the factory default settings in their software. That is like getting a briefcase with the tumblers on it, and leaving zero, zero, zero, zero in as the combination, and then locking your family jewels in it and expecting nobody to steal them. The default number never has been changed on one-third of all the installed software. I could probably pick most of your telephones, as well, by the way. Most of you probably never changed the default settings on your answering machines either. Most of you have one or two digit codes, and if I were patient enough, built a little program, I could dial into your homes and listen to all of your messages.

Ray Mellado: Second question, if I may, on information technology. How does it fit into K through 16 education?

Nick Donofrio: I don't have time to give a presentation on the future of technology, but let me just say that in 30 years we have experienced a six order of magnitude improvement—exponential or superexponential improvement. And it is going to continue that way for the next 30 years. All of that technology is likely to come to the "front of the house," meaning it is going to be where people use it.

I think the issue of computer literacy will eventually disappear because everybody is going to be naturally dealt with at the computer front end. I think computer literacy is terribly important, but the concept will be eradicated. Think about it. Are you electricity literate? Are you telephone literate? Are you tv literate? Do you understand the inner workings of those technologies? The answer is that you are not now, but 50, 60, 70, 80 years ago, you were. You were either electricity literate or you electrocuted yourself when you flipped the switch on. You were also telephone literate. It was either that or you listened to everybody's conversations, because most telephone connections were party lines. Computer literacy will disappear because natural interfaces will develop. Our young people will be able to use this stuff just like they use pencils and pens. We believe in a "pervasive" world, pervasive computing. The desk top era is over. We are going to have a computer-pervasive world, and human-friendly instruments and tools will be given to people.

That doesn't mean that the world will be less complicated or that we won't need any more big ideas. It means exactly the opposite. Remember, the simplification I talked about? We have got to find a way of simplifying the complexities we have created. That is where the real value will be. The need for information technology will be incredibly higher, and we will be able to use it more in teaching, not just in college and universities, but also in K through 12.

Participant: I would like to ask you a short question about your talk, which I really enjoyed. You did something I find myself doing, which is—and everybody in technology does it—making fun of people who go into management because they aren't doing real work. Do you think we are making our workplaces less culturally hospitable than they could be because we don't value management?

Nick Donofrio: Actually, that is an interesting question. In my company, we have the opposite problem. Our women and underrepresented technical people actually move to management far too often because they think it is easier. They think they can get ahead faster, and they don't want to compete on the raw bona fides of their intelligence and their capability. They think life will be easier for them, and in some cases they are right. I mean, some parts of our business are incredibly baroque, complex, difficult, and challenging.

Women are incredibly smart at finding the path of least resistance and getting there. They move very quickly to where they think they are going to rise to the top. I want to keep more women in technology longer. Of course, they can make up their minds and go wherever they want, but I want to see them go into management later, not earlier in life. The same thing is true for underrepresented minorities.

I am a manager, and I am an executive. Unfortunately, I am on the dark side, not the enlightened side. I mean, I have run $20 billion businesses, so I know what it is like on the management side. Management is important, but only if you have the technical talent to pull it off. You have to recognize that there are fewer management opportunities in a company than technical opportunities. We have 55 fellows, 210 distinguished engineers. Technical excellence is a better vehicle for technical people to become executives than for them to try to bend themselves into something else and climb the management ladder. We hear from technical people that they are being forced up the management ladder and that they don't like it, that they would prefer to have careers on the technical side.

We may be unique at IBM, and that may be just our problem. But I don't think so. I worry deeply about IBM because of our history—you have to remember that we started as a sales company. Mr. Watson, Sr., was a cash register salesman for NCR before he founded IBM. The company still puts a premium on selling and selling talents. I am trying to exert as much pressure against that as I can to keep people on the technical side as long as I can.

Summaries of Final Breakout Sessions

Where Do We Go from Here?

Following Mr. Donofrio's remarks, workshop attendees were again asked to participate in breakout discussions on the topic of where we go from here. The specific questions were: What obstacles remain? What research should be done? What are the policy implications?

Group 1 Summary

Cori Lathan
FIRST Robotics Competition and AnthroTronix

First, to get more young people excited about engineering careers, we need to publicize diverse engineering teams doing great things, and we need to show teacher successes—how great teaching can be and how successful it can be. We also need to publicize the best-practice programs in education and in industry.

Second, we want the NAE to establish a new award promoting K through 16 education, or maybe two or three awards, one for K through 6, one for grades 6 through 12, and one for 12 through 16. We need some awards, and we also need to increase the visibility of, and appreciation of, teachers.

Finally, homeland defense could be our next space race model. Effective homeland defense will require engineers. We need to communicate this need to our national leadership and make it clear that we can't have engineers without strong K through 16 education and good teachers (with increased pay and increased respect).

We have talked a lot about issues that have been talked about ad nauseam before, such as the image of engineers in our society, suggested improvements, best practices. Our question is how all of this can be incorporated into the workshop report. We don't know how to capture best practices and important suggestions in one report.

Peggy Layne (National Academy of Engineering): I can tell you that the NAE has established a new award, the Gordon Prize for Innovation in Engineering and Technology Education, which will be presented for the first time in 2002. The award recognizes an inspirational, innovative educator in engineering. The Gordon Prize addresses the 12 through 16 segment, but doesn't apply to pre-college education.

Group 2 Summary

Arline Easley
U.S. Department of Labor

In our group, we talked about overarching themes. We know we don't do a good job of tracking children in K through 12 into engineering. One of the problems may be that K through 12 teachers do not have a good idea of what engineering is and why it is important. Most of them don't understand that everything that makes our lives good has been developed by engineers. We believe that the NAE Committee on Diversity in the Engineering Workforce could play a pivotal role in carrying out our ideas by focusing on educating teachers and developing workshops or modules for teachers in K through 12, particularly grades 4, 5, 6, and 7, to show them how exciting engineering can be and how rewarding and exciting being an engineer can be. A related problem is counselors in the schools, who are educated in the same programs as teachers. Counselors are supposed to help students choose their career paths, but very often they have little impact. The NAE committee could focus its efforts on counselors, as well as teachers, educating them so that they can educate children. They could go a long way toward motivating women and minorities to enter engineering schools.

One of our members suggested that we treat the whole issue as an engineering problem and set goals, decide on deliverables, and break down implementation strategies. The NAE committee might be able to follow up on our IBM speaker's suggestion about consolidating some of the groups representing minorities to work together and send a single message rather than a cacophony of different messages.

There is some question about whether or not engineering faculties themselves are well grounded in educational processes. They may need a checklist of ways to encourage diversity in the classroom and responses to students of diverse backgrounds. Do they respond differently to white males than to women or minorities? The classroom can provide an environment in which students know they are welcome. We recognize that most educators have been getting this kind of information for 20 years, but the faculty in engineering schools may not be getting it. That may be a gap we can fill.

Finally, as a society, we may have to provide funds for engineering students, using the public health model—if you want to pursue a particular kind of education, then you have to work in the Public Health Service for a few years afterwards. That model might be very helpful for inner-city students

Group 3 Summary

Dundee Holt
National Action Council for Minorities in Engineering

We started by asking about research needs, but first we decided to take a step back and talk about the underlying issues. First, a member of the group noted that the public will to ensure educational equity does not exist; the public, Joe and Jane Average, are not really involved and don't have the will to make changes. Someone else pointed out that no strategies for change had been offered. The problem of educational equity seems so vast that people just say, oh, my, it is like the bogeyman. I don't want to deal with this. We need to set boundaries for the problem, define it, and present it in a way that makes Joe and Jane Average believe that they can do something about it. Part of our frustration in the NAE Diversity Forum, and even in these meetings, is that there is so much to be done that we don't know where to start. We decided to limit our discussions to things that are doable, that we could actually start on.

Another topic was the preparation of teachers. In his opening remarks yesterday, NAE President Wulf suggested that people find engineering repugnant, but I am not convinced of that. It is just that people don't have a clear idea of what engineering is because we haven't made it real to them. That is what our strategies and actions must do. To that end, we decided, after much discussion, that we should take a long-term approach to the problem, at the same time recognizing that companies will want to see results quarter by quarter. Some things will have to be done as we go along, but overall, we should adopt a long-term approach. The discussion focused on K through 12 education because, as everyone knows, if we don't fix that, we will be talking about these same issues 10 and 20 years down the road. As Nick Donofrio said this morning, not much has changed in the last 30 years. To make changes, we have to start with K through 12 education.

There are two issues related to K through 12 education that people on the outside would love to separate; but we believe they cannot be separated. We must not force people or ask people or encourage people or allow people to decide they are going to advocate for one or the other. If we don't address both issues, we will not succeed. If a child does not leave the fourth grade with proper reading and comprehension skills, there is no way he or she will be able to do the middle school math that will lay the foundation for more complicated material. As advocates for K through 12 education, we must focus on reading and comprehension skills in preschool through fourth grade, and then, in middle school, when the students already have a good educational foundation and are already excited about learning, we must focus on the certification of math and science teachers who can propel students on to the upper grades. We want corporations in particular to be advocates for, and to allocate their resources in both of these areas.

We talked about three specific actions we want corporations, government agencies, and educators to take. This morning, we were encouraged to bring our minority organizations together as one. Recognizing the need for national teacher certification and that many school districts do not have enough resources, we would like to see corporate America come together as one to develop a national institute for teacher certification in math and science. This institute would not belong to any particular corporation but would be supported by all of the corporations that have committed dollars and other resources.

Second, we want to leverage the power of the Internet. Many of you will remember back in the late 1970s or early 1980s when *Channel 1* first brought educational television to the classroom. We want a virtual presence that would bring engineering into the middle school classroom, a sort of virtual-age *Channel 1*. The kids and teachers could use it however they wanted. We could, perhaps, build some curricular material around it, but mostly it would be a resource for teachers who simply don't have the information they need or the resources to get it. By forming virtual partnerships with them, we would provide those resources. We believe the Internet offers a way to do that. We also want companies to move away from sending individuals into the classroom to talk to 30 or 40 students at a time and then patting themselves on the back. We want to move away from that kind of thing because it does not address the systemic problem.

Our third suggestion is the most out-of-the-box, radical suggestion. The context for this idea is that in 2000 minorities comprised a smaller portion of the freshman engineering class than they did in 1992, which means that in five years they are going to comprise a smaller portion of the graduating class. At the same time, we are trying to double the number of minorities in science and engineering. Last night, our dinner speaker, Congresswoman Eddie Bernice Johnson, suggested that we declare a national state of emergency and that we draft engineers into the nation's service the same way we drafted young men into the armed services. If a child has strong math and science skills and we believe that he or she could be an engineer, then we would "draft" that person. We would make sure he or she understood that being an engineer would be in the national interest. Mary Mattis of Catalyst mentioned that they had long thought of the shortage of engineers as a national security issue, and I have heard other people say that this idea came home to them on September 11. This is a national security issue, and we have to get it onto the national radar screen. We tried to look at the problem from 30,000 feet up rather than from the ground. We tried to stay away from details, but we know we will have to offer incentives. If we are going to encourage a young person to study engineering, we have to provide some incentives.

Another thing that can be done is for companies to commit themselves to lifelong training. The fact is that not enough people are graduating with engineering degrees. There are, however, enough people, or at least many more people, who graduate with the skills to become engineers. Joel Harmon of Shell Oil said that they have a lot of chemists, perhaps too many chemists (if one can ever have too many chemists). Shell brings those people on board and after their first year trains them to be chemical engineers. Those kinds of human resource strategies could take us a long way toward meeting our human resource needs.

Closing Address

The Business Case for Diversity

JAMES J. PADILLA[1]
Group Vice President, Ford North America
Ford Motor Company

I've been invited to discuss the business case for diversity—why diversity is as critical to your business strategy as the products you make or the services you provide—and to discuss the benefits for companies that not only seek but celebrate diversity in their offices and plants, as well as in the ideas that shape their companies and their products.

It's an easy case for me to argue. As a group vice president at Ford and a member of Ford's Executive Council on Diversity and Worklife, I am invested in the value of creating a diverse culture in our company. As the grandson of immigrants, I also have a deep personal interest in the issues of diversity, inclusion, and justice. I'd like to begin by talking briefly about our diversity journey at Ford—where we've been and where we're going as we near our 100th anniversary as a company. Like most leading companies, we view the twin concepts of diversity and inclusion as critical to our future success. Many people do not realize, however, just how important these concepts have been to our past.

Nearly a century ago, our founder, Henry Ford, was among the first to cultivate a workforce from all of the communities the company served. He opened his plants, offices, trade schools, and supervisory ranks to minorities decades before other manufacturers. His grandson, Henry Ford II, built upon this foundation. He championed providing access and opportunities for all people, inside and outside the company. The Ford family's commitment to social responsibility continues today with our current chairman Bill Ford, Henry Ford's great grandson.

---

[1]Mr. Padilla's remarks are used with permission.

I'm proud to say that Ford now has the largest number of minority dealers in the country, with more African-American dealers than all other automakers combined. We purchase more goods and services from minority suppliers in the United States than any other corporation in the world. In recent years, we have strengthened our support of minority organizations in the communities we serve. We have expanded charitable donations in the areas of education, health and welfare, and arts and humanities with a strong emphasis on programs that promote diversity. We have provided scholarships, internships, and financial support to colleges and universities. And we have worked closely with community leaders to find new ways to make a difference.

We clearly are headed in the right direction. But more important, we've expanded our notion of what diversity means in a global corporation today. At Ford, our definition of diversity goes far beyond what we look like or where we are from. It includes all of the traits that make us unique individuals. It also includes the way each of us works, as well as how we choose to blend our professional and personal lives.

As company leaders, we are challenged daily to be flexible and to recognize that there are as many ways to do a job as there are people in the world. True diversity, I believe, celebrates the rich qualities and experiences employees bring to their jobs each day and considers those qualities to be among the company's greatest assets. This, of course, is a matter of fairness and justice, but it is also good business. We are a global organization with employees and customers around the world. Therefore, we must understand our business from our customers' perspectives—and those perspective are becoming increasingly diverse.

To fully appreciate these diverse perspectives and to ensure that we have the best talent available to do so, we believe it is critical that we create a culture of inclusion at Ford. In this culture, every employee is welcomed, supported, respected, and encouraged to make a contribution and to be successful. We want diversity to be in the bloodstream of our company. That is a promise we have made to every member of our global family.

Diversity is about much more than policies and programs. It's a commitment that starts at the very top. From the way we design our automobiles to the way we market our products, we strive to include all perspectives. That's because we believe an inclusive environment is a formula for lasting success for a business... a community...even a nation. Voices that are silenced or ignored, for whatever reason, represent not only an injustice, but also a valuable resource that has been wasted, a tragic waste of human capital. Enlightened corporations understand that these are important business issues. They realize they cannot separate themselves from what is going on around them. They realize that, ultimately, they can only be as successful as the communities, and the world, in which they exist. Therefore, they must reflect that world.

Only a diverse company can fully understand our diverse and complex global marketplace. Only a diverse company can generate breakthrough ideas that will lead to the development of innovative products and services—products and services that will meet the diverse needs of our customers. I've often said that if we want the world to buy our products, we must think like the world—and see the world through the eyes of its many peoples. If we fail to reflect a wide variety of viewpoints and experiences in the products we design, we will lose the opportunity to serve critical markets—powerful, growing markets that research tells us are essential to future success.

Have we reached this ideal at Ford? Not yet. It's one thing to say we support diversity—quite another to transform the corporate culture in a global company as vast as ours. But we are making progress. In the past several years, we have reengineered our processes and policies to incorporate these values into our business plans. Momentum will continue to build as we continue the hard work necessary to achieve true and lasting cultural change. Each and every day, this means we must translate our values into practice and bring them alive for the people in the company—as well as for our customers, our communities, our dealers and suppliers.

We must work every day to develop solutions and new strategies and to explore new ways of doing things. Throughout our company, we must seek to create workplaces that are fair, open, and inclusive, where all employees have the opportunity to realize their full potential based on their skills and merits and where they are supported in the fulfillment of the many roles they play in their lives—as professionals, parents, partners, spouses, sons or daughters, volunteers, community activists, and students.

The Executive Council on Diversity and Worklife, a 26-member executive committee is leading this effort. Working in partnership with the Corporate Diversity and Worklife Office, we oversee diversity efforts at Ford worldwide by providing strategic direction and programs that are aligned with other initiatives. The company has also created a Global Diversity Council to help develop and implement programs on a worldwide basis. This has included the appointment of diversity managers in key global markets who are in the best position to understand the needs and issues of local cultures and communities. The Global Diversity Council also actively benchmarks Ford's performance against the performance of other companies and shares best practices from around the world. Nearly every Ford location and every Ford plant has a local diversity council.

All of this hard work reflects a wide range of activities related to diversity initiatives. At the Global Diversity and Worklife Summit this fall, we discussed refinements to recruiting and mentoring programs and new communication tools—including a global diversity and worklife brochure and Web site. Because we strive to make all employees feel comfortable and valued at work, my organization—manufacturing—has also taken steps to develop antiharassment training for all of our U.S. plant employees. This training began as part of an EEOC settlement in Chicago, and we have voluntarily expanded it to include all of our plants.

Early studies suggest that antiharassment training has been very effective in raising employee awareness of what harassment looks like. It also has clearly defined the responsibilities of each manager and employee in maintaining a fair and respectful plant environment. Our company also works closely with nine employee resource groups—we call them ERGs—to give all employees a voice in the future of Ford. The ERGs are company-recognized organizations formed by employees with common interests, backgrounds, or perspectives. Each ERG, which represents an important employee group in our company, is championed by a senior executive.

I'm pleased to be the executive liaison for the Ford Hispanic Network Group, an extremely active ERG that was formed by my Uncle Leo, a retired Ford employee, in 1992. Because I know it best, I'd like to use the Ford Hispanic Network Group—the FHNG—as an example of how Ford and its employees are working together, not only to promote diversity, but also to leverage the company's power to touch our consumers and their communities.

The Hispanic population in the United States is growing phenomenally, far exceeding census projections. In fact, tortillas now outsell bread in the United States! This is our new America, folks. Now let's take a deeper look at the demographics that are changing the ethnic landscape of our country. Hispanic purchasing power is expected to double every ten years, reaching an astounding $2 trillion by 2020. At 35 million, the Hispanic population now outnumbers the population of Canada by five million people. Today, one in three people living in California and Texas is Hispanic. Hispanics also represent a young market—40 percent are under 20. The population of U.S. Hispanics in households with median incomes increased by 70 percent from 1980 to 2000, compared with a 14 percent increase in the U.S. as a whole. Approximately 20 percent of U.S. population growth in the next 10 years will come from the Hispanic segment. Our retail sales to Hispanics represented more than $3.6 billion in 2000—a 45 percent increase since 1998.

It's easy to understand why companies are interested in understanding and connecting with this lucrative market. We all know this is the only way we can create products and services that interest Hispanics. The challenge is to create strategies that target this specific market—comprehensive strategies that will ultimately influence everything from our hiring practices to product development and marketing.

I'd like to share with you what we have done at Ford, in close partnership with the FHNG. Our first area of focus has been on recruiting and development. Although the representation of Hispanics in our workforce is increasing, we know we must do better to keep pace with the changing population. Since 1998, we have nearly tripled the number of colleges and universities with high Hispanic populations where we actively recruit. We've expanded our scholarship programs as well. The FHNG and Ford are working together on these initiatives. Community outreach is another critical component of our strategy. In this area, our partnership with the FHNG is invaluable. I have worked side by side with group members on a number of volunteer efforts, including cleanup of a local park in a predominately Hispanic neighborhood.

I'm very excited about the potential of our largest outreach project, the Southwest Detroit High School Partnership Program. As part of this program, Ford employees actively mentor, provide tutoring, and hold seminars on job interviewing skills and other topics related to business success for at-risk kids in Detroit high schools. The program is a joint effort of the company, the FHNG, and the Ford African-American Network. The goal is to establish a real presence in the Detroit Hispanic and African-American communities.

We cannot accomplish our goals without strong support from our employees. That is why our third strategy is to strengthen and expand the Hispanic network at Ford. I'm pleased to say that we've had great success. In 1998, there were only 98 FHNG members. Today, there are more than 500, and nine satellite chapters will be added this year.

As I mentioned before, the FHNG is only one of nine resource groups, each of which helps us in a similar way to understand our markets, improve our recruiting efforts, and connect with the communities we serve. We often ask ERGs for advice on product development, marketing strategies, and business plans so we can be in touch with our customers and meet their needs with new products and services.

I'm certain you will agree that these are essential steps for any company that strives to be consumer-driven. At Ford, no matter who we are, where we live, or where we work in the company, we are united by common threads. We all want Ford to be a profitable, customer-focused, global company. I believe a diverse workforce and a flexible work environment are keys to business success for a company like ours. In our industry, all of the main players have similar strategies, products, and services, so the competitive advantage must be in intellectual capital—the way our people tackle challenges with innovative and creative thinking. If we can bring together diverse minds and apply them to automotive challenges, we'll have an amazing kaleidoscope of ideas to choose from, breakthrough ideas to meet the needs of our customers in a complex and increasingly diverse world.

Questions and Answers

Lisa Gutierrez (Los Alamos National Laboratory): Can you tell us more about the Earn to Learn Program? Where can I get a sample of that curriculum so we can see how we might roll it out in New Mexico?

Jim Padilla: Focus Hope has a well documented program and a Web site, *www.focushope.edu*. The program has been in existence for some 30 years and is becoming a national model of how to do these things. If that doesn't work, call me, and I will get you some information about how the programs work.

Dundee Holt (NACME): You talked about the difficulty of persuading a kid from San Antonio to come to Detroit. One of the things that keeps us up nights is how to get companies to go where the students are. California is the largest producer of minority engineers, but a lot of companies don't recruit there because they figure they can't get somebody from San Francisco to come east. What does Ford do? What would you recommend for other companies?

Jim Padilla: You have to set realistic expectations, because it is a tough sell. We have one significant advantage, in my view. Everybody can identify and likes our products, and they want to be part of that. But I think you have to expand your recruiting net. You have to be consistent and persistent over time to build relationships. We are in a compression in the economy right now, which is going to make things more difficult. We are going to have to taper back our overall recruiting. That is a fact of life. But we still have to do the right things to make sure that our workforce is inclusive.

We will continue to go to universities and develop relationships and seek out the best students we can find. Our biggest challenge—we have done very well with females—is finding minorities. It is very, very difficult. That is why I am convinced we should consider "growing" our own. That is why I want to continue to work in southwest Detroit. I won't have to convince those kids to come to Detroit.

We can encourage kids to go to school in the local area by providing them with scholarship money and with internships. Let me tell you about what we did last summer. We provided about 40 internships for high school students from various schools in the area and then we had a luncheon for them. At the luncheon we asked each one to stand up and give a three-minute summary of what they had learned and what they planned to do next. A lot of these kids had never been on the stump before, and they did a great job. One thing that impressed me is that we created in their minds an elevated line of sight, which is very important. They realized they could do an important job and be involved in something big. We also appointed mentors to help usher them through. The kids did a good job, and I was very proud of them.

Theopolis Holeman (Duke Energy): Your passion is evident and contagious. I am curious as to how you nurture that passion, particularly for some of your counterparts in management. You, obviously, are an advocate, but I am curious about whether that is the culture at Ford or whether you have to educate others to bring them along with you.

Jim Padilla: I think you always have to educate others. You have to make sure that the network groups, for example, are "plugged in" to the organization and have good sponsorship and that the sponsor meets regularly with their champion. I have breakfast about every six or eight weeks with some of the network group leaders to listen to them and figure out how I can help. I also bring other people in to help groups find their way. You will find that people really do want to be involved.

Sandra Begay-Campbell (NAE Committee on Diversity in the Engineering Workforce and Sandia National Laboratories): When we were preparing this conference and deciding who to invite, we read many interesting articles, some on controversial issues, such as backlash among groups opposed to or offended by resource networks. A couple of articles were about white males joining together against Ford because of Ford's emphasis on resource networks. Can you talk a little bit about backlash?

Jim Padilla: I don't think the backlash is really against the network groups. I think the issue is that every group feels like a besieged species. You have to sift through the responses to find the real problems. I prefer dealing with things like engineering projects, rather than litigation. Somehow, we have to convey a message that we are not giving anyone special privileges, but that we are trying to provide opportunities for everyone.

Sometimes we don't convey that message as well as we should. Our representation has improved dramatically, but, frankly, it is still way behind for Hispanics, for example. The population of Hispanics is going to be huge, and we are underrepresented. Other minorities are also underrepresented, particularly in the higher levels of management. Therefore, we have to fill the pipeline with a diverse group of candidates. We are trying to make sure we have broad representation from different populations, which is bound to create some controversy. If you are aggressive in these policies, you are bound to raise some issues, but I think that you can get through those. We can find ways to be more sensitive. Frankly, we probably lack sensitivity in some of these areas.

The last thing we want to do is alienate important groups of people in our company, because everybody has a role to play. I am sure there will be litigation, and I am sure some interesting insights will come from the courts. But in the end, this is a hearts-and-minds game. The more we can involve everyone and the better people understand that this is the right way to go, the better it will be for our business and for our customers.

Gary Downey (Virginia Tech): What kind of resistance have you encountered in the organization, and how has resistance been articulated? How do you deal with resistance without polarizing the conflict and creating entrenched camps?

Jim Padilla: Over time I think that the merits and capabilities of the individuals you move into positions will be evident. They may not be immediately evident, and there are some risks. Some of the things we do are a stretch. That's why it is important to have support mechanisms in place to make sure that the stretch isn't too long. In some cases, you need special supports.

You have to do this with a good degree of sensitivity, and we are learning from our mistakes. But in my view, this doesn't change the course we are on. You have to do things with sensitivity, and you can't expect everybody to stand up and do handstands and cheers. The courts may not always support you either.

Joan Straumanis (U.S. Department of Education, Fund for the Improvement of Post Secondary Education): Something remarkable happened today. The breakout session reports weren't like the ones we have heard in previous conferences. I have been a member of the NAE forum for a couple of years, and this is the first time I have seen a real focus on K through 12 teachers. In fact, all three groups focused on teachers, which was not part of the assignment. It wasn't even in the air, as far as I could tell, and yet, everybody came to the same conclusion. That was very gratifying to me, because I think improving teachers and teaching in K through 12 is the No. 1 problem in this country.

You talked about Ford's working with kids in the Detroit schools. Are you working with teachers and teacher education, at Detroit-Mercy, for example? I suggest that you focus some of your attention there, especially in terms of engineering literacy. It is not just teachers' skills we have to change, but also attitudes.

Jim Padilla: That is a very good question, and I don't think we have adequately addressed it. As we involve more schools, in some of the robotics programs, for example, teachers naturally come. We also participate with the university through sponsorships and so on, but we need to do more.

As we anticipate a shrinking workforce in the automotive industry, one of the things we are working on now, is encouraging technically competent employees who opt for voluntary retirement programs to consider teaching. We offer very generous separation packages for these people so they may not have to race out and find another job. We are looking for individuals who can make a commitment to teaching, and we will see how that works. We are working with the dean of engineering at the University of Detroit on that. But let's be honest. The economic drivers for people competent in math and sciences are to go to an information technology company or someplace where they can make more money.

Karl Pister (University of California): I applaud what you are doing in Detroit, and I hope you can attract more people from California because we are getting overpopulated. You mentioned going into a school to fix an out-of-date laboratory, and certainly one must applaud that. But what about all the schools in Detroit you don't reach where laboratories also need to be brought up to date? In other words, doesn't Ford, and other corporations in Detroit, as well as the public have an obligation to see that all of the high schools in Detroit have up-to-date equipment? Wouldn't that be a better place for Ford to use its political might?

Jim Padilla: I think we have a very good record of donating from the Ford Motor Company Foundation. We spend or donate millions of dollars for a lot of the institutions represented here today.

Karl Pister: It is not the money. It is the desire or the obligation, the opportunity, to restore our public education system to a level of equality.

Jim Padilla: That is a good point. From my perspective, you have to pick some spots and make a difference, set up some role models. Companies aren't just cash machines—Ford is losing money hand over fist right now. Our funds are very limited, so we have to pick our spots. We have chosen to work primarily with about 30 universities. We focus on programs we think are important, not only for the development of technical skills, but also for emerging technical arenas, like the environment. We just set up an environmental center at Georgia Tech.

I agree wholeheartedly that we don't do enough in secondary education or in K through 8. I think that will require a broader forum and broader support because no single industry or company or even government body has the resources and the wherewithal to address the issues of entire cities and states. It is an important topic, though, and one of the recommendations of this gathering ought to be forming that type of coalition. By the way, I come from a very large family. I have eight brothers and two sisters. One of my brothers is a science education teacher at Georgia Tech, and he used to be the head of their science education department. He lectures me all the time about my responsibilities, so I appreciate your reminder.

Appendixes

Appendix A

Workshop Participants

Arnold Allemang
Vice President, Operations
Dow Chemical Company

Daniel E. Arvizu
Group Vice President
CH2M HILL

Ray Beebe
Consultant

Sandra Begay-Campbell
(Member, NAE Committee on Diversity)
Senior Member of the Technical Staff
Sandia National Laboratories

Lara Black
Human Resources Manager
TRW Systems

Mike Bober
Recruitment Coordinator
ExxonMobil Corporation

Suzanne Brainard
(Member, NAE Committee on Diversity)
Director, Center for Workforce Development
University of Washington

Mary Cleave
Deputy Associate Administrator
NASA Office of Earth Sciences

Toni Clewell
Principal Research Associate
Urban Institute

Richard P. Cowie
Vice President, Human Resources
Consolidated Edison

Jose Cruz
(Member, NAE Committee on Diversity)
Howard D. Winbigler Chair in Engineering
Ohio State University

Lance Davis
Executive Director
National Academy of Engineering

Nicholas Donofrio
Senior Vice President, Technology
and Manufacturing
IBM Corporation

Gary Downey
Center for Science and
Technology Studies
Virginia Polytechnic Institute and
State University

Arline Easley
Economist
U.S. Department of Labor

Jose Font
Operations Senior Manager
The Boeing Company

Janie Fouke
Dean, College of Engineering
Michigan State University

Lenny Fraser
Diversity Consultant
Career Communications Group, Inc.

William Friend
*(Member, NAE Committee on
Diversity)*
Executive Vice President (retired)
Bechtel Group, Inc.

Janet M. Graham
Training and Development Specialist
DuPont Human Resources

Domenico Grasso
Founding Chair, Picker
Engineering Program
Smith College

Lisa Gutierrez
Diversity Director
Los Alamos National Laboratory

Orlando A. Gutierrez
Consortium Chair
Society of Hispanic Professional
Engineers

Joel Harmon
Manager of Diversity and
Work/Family
Shell Chemical LP

Margaret Harvey
Diversity, EEO & AA Services
Sandia National Laboratories

Theopolis Holeman
Senior Vice President, Transmission
and Engineering
Duke Energy Gas Transmission

Dundee Holt
Vice President, Public Information
National Action Council for
Minorities in Engineering

Rachel Ivie
Senior Research Associate
American Institute of Physics

J.J. Jackson
Executive Assistant to the President
Babson College

Brenda Jackson
Executive Vice President,
 Business Services
Texas Utilities

James Johnson
Dean, School of Engineering
Howard University

Arnold Kee
Coordinator of Minority Services
American Association of
 Community Colleges

Sarah Ann Kerr
Investment Engineer
DuPont Engineering

Klod Kokini
Assistant Dean of Engineering
Purdue University

Ralph Larson
Staff Vice President
3M Engineering

Cathy Lasser
(Member, NAE Committee on Diversity)
Vice President, B2B Initiatives
IBM Corporation

Corinna E. Lathan
Special Projects Advisor
FIRST Robotics Competition
President and CEO, AnthroTronix

Peggy Layne
Staff Officer
National Academy of Engineering

Michele Lezama
Executive Director
National Society of Black Engineers

Mary C. Mattis
Vice President
Research and Advisory Services
Catalyst

Raymond G. Mellado
Chairman and CEO
Hispanic Engineer National
 Achievement Awards
 Corporation

Rodney Nathan
Global Diversity Manager
Duke Energy

Lisa Nungesser
Senior Vice President
Parsons Brinckerhoff Quade &
 Douglass

James J. Padilla
Group Vice President, Ford
 North America
Ford Motor Company

Kimberly Patterson
Senior Vice President of
 Human Resources
BE&K

Willie Pearson, Jr.
Professor and Chair, School of
 History, Technology and
 Society
Georgia Institute of Technology

Carrie Phillips
Governmental Relations Manager
Southern Nuclear Operating Company

Karl Pister
Chancellor Emeritus
UC Santa Cruz

David M. Porter, Jr.
(Member, NAE Committee on Diversity)
Assistant Professor, Human Resources and Organizational Behavior
UCLA

Frank Quevedo
Vice President, Equal Opportunity
Southern California Edison

Cordell Reed
(Chair, NAE Committee on Diversity)
Senior Vice President (retired)
Commonwealth Edison Company

James Rosser
President
California State University, Los Angeles

Helmut Schuster
Human Resources Manager, Global Refining
British Petroleum

Tyra Simpkins
Diversity Consultant
Career Communications Group, Inc.

Robert Spitzer
Vice President, Technical Affiliations
The Boeing Company

Susan Staffin Metz
Executive Director, The Lore-El Center for Women in Engineering and Science
Stevens Institute of Technology

Joan Straumanis
Program Officer
U.S. Department of Education

Richard Taber
Corporate/Foundation Relations Consultant
National Science Foundation

Tyrone D. Taborn
CEO
Career Communications

Richard Tapia
Noah Harding Professor of Computational and Applied Mathematics
Rice University

Iwona Turlik
Corporate Vice President and Director
Motorola Advanced Technology Center

Dale Von Haase
Director, Aerospace Sciences
Lockheed Martin

Wanda E. Ward
Chief Advisor to the BEST Initiative
Council on Competitiveness

Ming-Ying Wei
Earth Science Education Program
 Manager
NASA Headquarters

James West
*(Member, NAE Committee on
 Diversity)*
Distinguished Member of the
 Technical Staff and Fellow
Bell Laboratories
Lucent Technologies

Kurt Wiese
Downstream Human
 Resources Manager
ExxonMobil Corporation

Thomas S. Williamson, Jr.
Partner
Covington & Burling

Shelley A.M. Wolff
President
Society of Women Engineers
HNTB Corporation

Jeff Wright
Dean of Engineering
University of California Merced

Wm. A. Wulf
President
National Academy of Engineering

Linda Young
Senior Human Resources Manager,
 Policy, Diversity, Retention,
 Employee Relations and
 Compliance
TRW Systems

Appendix B

Biographical Sketches of Workshop Speakers

Daniel E. Arvizu is senior vice president and technology fellow in the CH2M HILL Energy, Environment, and Systems Business Group. Prior to joining CH2M HILL, Dr. Arvizu worked for 21 years at Sandia National Laboratories, where he managed leading-edge research in areas ranging from solar energy to nuclear weapons. He also led the development of industry/laboratory partnerships for technology commercialization and has been actively involved in supporting university programs to prepare future scientists and engineers for the workforce. Dr. Arvizu presently serves on the U.S. Department of Energy National Coal Council and U.S. Department of Defense Army Science Board; he is an advisor to the National Research Council Division on Engineering and Physical Sciences. In 1996, he was awarded the Hispanic Engineer National Achievement Award for Executive Excellence. He received his B.S. in mechanical engineering from New Mexico State University and an M.S. and Ph.D., also in mechanical engineering, from Stanford University.

Sandra Begay-Campbell, a Navajo, is a senior member of the technical staff at Sandia National Laboratories and board chair and executive director emerita of the American Indian Science and Engineering Society (AISES). Prior to working at Sandia, Ms. Begay-Campbell worked at Los Alamos National Laboratories and Lawrence Livermore National Laboratory. She received an M.S. in structural engineering from Stanford University and a B.S. from the University of New Mexico.

Richard P. Cowie was elected vice president of employee relations (now called human resources) for Consolidated Edison in March 1994. He joined the company in 1963 and has held positions of increasing responsibility. From 1980 to 1986 as general manager of Manhattan customer service, he led the company's efforts to decentralize into geographic branches. He was named director of credit and collections in 1986 and assistant to the executive vice president of division operations from 1988 to 1990. In 1991, he was named director of customer service. He is also a member of the Board of Trustees of the American Red Cross of Greater New York and the Board of Directors of the Fourteenth Street-Union Square Local Development Corporation. Mr. Cowie earned a B.S. in economics from the College of Staten Island and an M.B.A. from Columbia University.

Nicholas Donofrio leads the strategy for developing and commercializing advanced technology for IBM's global operations. His responsibilities include overseeing IBM research, the Global Integrated Supply Chain Team, the Integrated Product Development Team, and the Import Compliance Office. He also leads IBM's worldwide quality initiatives. He is chairman of IBM's Corporate Technology Council, chairman of the Board of Governors for the IBM Academy of Technology, a member of IBM's Corporate Development Committee, and a member of the IBM Chairman's Council. Mr. Donofrio spent the early part of his career in microprocessor development as a designer of logic and memory chips. Mr. Donofrio is a strong advocate of education, particularly in mathematics and science, the keys to economic competitiveness. His focus is on advancing educational, employment, and career opportunities for underrepresented minorities and women. He is a member of the Board of Directors for the National Action Council for Minorities in Engineering, a fellow of the Institute of Electrical and Electronics Engineers, and a member of the National Academy of Engineering. Mr. Donofrio earned a B.S. in electrical engineering from Rensselaer Polytechnic Institute and an M.S., also in electrical engineering, from Syracuse University. In 1999 he was awarded an honorary doctorate in engineering from Polytechnic University.

Janet M. Graham joined E.I. du Pont de Nemours in 1989, where she is a training and development specialist in the People and Organizational Development Group, Corporate Human Resources. Ms. Graham is a consultant in the company and to other organizations on mentoring and has helped several businesses establish mentoring programs specific to their business needs. Her clients include American Airlines, Bell Canada, Conoco, Greenwood Trust, and Michelin. Ms. Graham is a member of the Mentoring Institute, the International Mentoring Association and the DuPont Mentoring Excellence Committee. In 1998, she was cocoordinator of the DuPont Mentoring Conference. She is a frequent speaker at conferences on mentoring.

Orlando A. Gutierrez is a national past president of the Society of Hispanic Professional Engineers. He retired in 1992 after 31 years with the National

Aeronautics and Space Administration (NASA) where he was an aerospace research engineer for 21 years. He has published more than 25 papers on space power-generation systems and jet-noise suppression. In addition, he served for eight years as manager of NASA's Hispanic Employment Program and two years as manager of the Minority University Program. Mr. Gutierrez has been recognized by many organizations for his educational, recruitment, and community activities. He is the recipient of the Society of Hispanic Professional Engineers Jaime Oaxaca Award; the Equal Opportunity Medal and the Exceptional Service Medal, both from NASA; and the Medalla de Oro from the Society of Mexican American Engineers and Scientists.

Sarah Ann Kerr, who began her career at E.I. du Pont de Nemours in 1995, is an investment engineer responsible for estimating and controlling costs for capital investments. Since 1999, she has been instrumental in the development of a mentoring program for the DuPont engineering organization. She coordinated the design and implementation of a training workshop and has been active in all phases of the mentoring program. Ms. Kerr has made numerous presentations on mentoring to DuPont organizations and external groups.

Michele Lezama, who was an active volunteer leader for 12 years, is now the executive director of the National Society of Black Engineers (NSBE). She oversees the operations of the NSBE World Headquarters staff and is chief operating officer of this student-run organization of 15,000 college students, technical professionals, and precollege students. Ms. Lezama also has experience at HBO, where she was director of scrambling operations; CBS Television, where she was associate director of broadcast operations; and IBM, Raytheon, and Texas Instruments, where she held engineering positions. At IBM, she volunteered to serve as the director of an employee outreach program called PREMISE (Poughkeepsie Regional Effort for Minority Introduction to Science and Engineering). Ms. Lezama has a B.S. in industrial engineering from Northeastern University and an M.S. in industrial engineering and an M.B.A. in finance and accounting, both from Columbia University. She is a member of Tau Beta Pi, Alpha Pi Mu, a GEM Fellow, and a Robert Toigo Fellow.

Mary C. Mattis is a senior research fellow at Catalyst, a nonprofit organization that works with corporations and professional firms to retain and advance women. Dr. Mattis conceptualizes and implements Catalyst's research on women's leadership development and is a frequently requested speaker on this topic. For 10 years, she led the Catalyst team that evaluates corporate gender diversity initiatives for the Catalyst Award. She is the author of the Catalyst publications, *Women in Engineering* and *Women Scientists in Industry*, as well as numerous other Catalyst research reports. She is regarded as an international expert on corporate best practices for retaining and advancing women and has published extensively on this subject. In 1999, she spent a month in Australia and New Zealand conducting workshops and speaking to business and government groups on gender diversity best practices. Dr. Mattis earned her Ph.D. in sociology from Washington University.

Lisa Nungesser is senior vice president, Parsons Brinckerhoff Quade & Douglass and vice president, Parsons Brinckerhoff Aviation, Inc. Her work has focused on applied research, management, and community involvement, and she has directed numerous multidisciplinary studies, including demographic and transportation studies, economic feasibility studies, and facility siting studies through environmental impact assessments and statements. Her experience and training include both legal and technical aspects of environmental management. She is Parsons Brinckerhoff's area expert on environmental justice and community impact, and has extensive experience in the National Environmental Policy Act process, including a technical background in socioeconomics, demographics, and citizen participation in infrastructure projects. Prior to joining Parsons Brinckerhoff, Dr. Nungesser was president and founding principal of a planning and research consulting firm where she focused primarily on community involvement and location theory. Earlier, as a transportation researcher for the Texas Transportation Institute of Texas A&M University, Dr. Nungesser conducted several socioeconomic forecasting studies in the lower Rio Grande Valley and highway corridor studies for San Antonio and McAllen, Texas. She has a Ph.D. in community/regional planning and geography from the University of Texas at Austin.

James J. Padilla, group vice president, Ford North America, is responsible for all operations, including manufacturing, product development, marketing, and sales of Ford cars and trucks in the United States, Canada, and Mexico. Mr. Padilla joined Ford in 1966 as a quality control engineer. In 1976, he was promoted to a series of management positions in product engineering and manufacturing. These included program operations manager for several car lines and director, Small Car Segment, Car Product Development. From 1992 to 1994, Mr. Padilla was director of engineering and manufacturing, Jaguar Cars, Ltd., during a critical turnaround period. He subsequently oversaw the successful launches of the Jaguar XJ series, the Jaguar XK-8, and the world-class AJ26 engine, the Aston-Martin DB-7, and the Lincoln LS. He earned a B.S. and M.S. in chemical engineering and an M.A. in economics from the University of Detroit. Mr. Padilla is a member of the National Academy of Engineering and was named Engineer of the Year by the Hispanic Engineer National Achievement Awards Conference.

Willie Pearson, Jr., is chair of the School of History, Technology and Society, Ivan Allen College, Georgia Institute of Technology. A specialist in the sociology of science and sociology of the family, Dr. Pearson is the author or coeditor of six books and monographs and numerous articles and chapters. His newest book in progress is *Beyond Small Numbers: Voices of African-American Ph.D. Chemists* (JAI Press, 2002). He has held research grants from the National Science Foundation, National Endowment for the Humanities, Sloan Foundation, and U.S. Department of Justice and postdoctoral fellowships at the Educational Testing Service and the Office of Technology Assessment. He is a lecturer in Sigma Xi's Distinguished Lectureship Program and chair of the Committee on Science, Engineering and Public Policy, American Association for the Advancement of Science. Dr. Pearson received his Ph.D. in sociology from Southern Illinois University (SIU) and the Alumni Achievement Award from SIU in 1993. He completed his undergraduate education at Wiley College.

Tyrone D. Taborn, chairman and CEO of Career Communications Group, Inc. (CCG), is the publisher and editor in chief of *US Black Engineer & Information Technology*, the only general-interest technology magazine for the African-American community. Mr. Taborn is also the producer of the award-winning syndicated TV show, "Success Through Education," and publisher of *Hispanic Engineer & Information Technology*. Mr. Taborn has been a guest editorial writer for the *Baltimore Sun* and writes a technology column that appears in newspapers, magazines, and on Web sites. He is currently a member of the National Association of Hispanic Journalists and the Baltimore Engineering Society and has served on the boards of directors of the Afro-American Newspaper Company, the Baltimore Urban League, and the Granville Academy.

Mr. Taborn is the founder of the Black Family Technology Week Program, which is sponsored by IBM Corporation, Microsoft, and Sun Microsystems. He was recently selected as an Internet and Technology Leader by Sprint and MOBE IT, which recognizes minority leaders. He was named one of the 50 Most Important African-Americans in Technology by the editors of Blackmoney.com and souloftechnology.net. An Afro-Latino who grew up in Los Angeles, Mr. Taborn attended Cornell University, where he majored in government. He was also a member of Quill and Dagger and the Telluride Association.

Iwona Turlik received her M.S. in electrical engineering and her Ph.D. in technical science from the Technical University of Wroclaw, Poland, where she started her professional career as a tenured faculty member. She has worked as manager of exploratory processing with Bell Northern Research, where she was involved in process and device development for several Si and III-V technologies. Dr. Turlik was director of advanced packaging technology programs, Microelectronics Center of North Carolina, and a tenured professor in the Electrical Engineering Department, University of North Carolina, Charlotte. She joined Motorola in 1994 as vice president and director of the Corporate Manufacturing Research Center and is currently corporate vice president and director of Motorola

Advanced Technology Center and corporate vice president and director of Motorola Technology Acquisition Office. She has published more than 100 professional papers and presentations, edited two books, and holds 13 patents. Dr. Turlik was named one of the 50 Most Influential People in the PCB industry by *PC FAB* and *ATOMIC29*. She is also a fellow of the Institute of Electrical and Electronic Engineers (IEEE) and a recipient of the 1994 Board of Governors Distinguished Service Award for outstanding contributions to the IEEE Components, Packaging, and Manufacturing Society.

Thomas S. Williamson, Jr. a partner at Covington & Burling in Washington, D.C., focuses on litigation and employment-related law. From 1993 to 1996 he was the solicitor of labor, the chief legal officer for the U.S. Department of Labor, where he oversaw a staff of 700, including 500 lawyers located in Washington, D.C., and in 15 regional and subregional offices. The Office of Solicitor is responsible for enforcing approximately 180 federal statutes related to wage protections, worker health and safety, pension benefits, employment discrimination and affirmative action, labor standards, job training, union democracy, unemployment insurance, and various other employee benefit programs. He represented the Labor Department as a member of the Board of Directors of the Overseas Private Investment Corporation and as a delegate to the International Labor Organization. He also participated in the interagency working group that negotiated the labor-related part of the North American Free Trade Agreement. Mr. Williamson earned his B.A. in social studies from Harvard University and his L.L.D. from the University of California, Berkeley.

Shelley A.M. Wolff is national president of the Society of Women Engineers (SWE) and associate vice president of HNTB Corporation, where she manages the highway design department in the Kansas City office; she is also director of corporate project management training programs and operations officer for HNTB's Corporate Business Services. Ms. Wolff leads SWE's outreach, education, and professional development programs. In 2002, SWE will focus on enhancing the image of engineers, one of the main recommendations of the report by the Commission on the Advancement of Women and Minorities in Science, Engineering and Technology. Ms. Wolff earned her B.S. in civil engineering from Iowa State University and her M.S. in engineering management from University of Kansas. She is a licensed professional engineer in Kansas, Iowa, Missouri, Nebraska, and Arkansas.

Wm. A. Wulf is president of the National Academy of Engineering and vice chair of the National Research Council, the principal operating arm of the National Academies. He is on leave from the University of Virginia, Charlottesville, where he is AT&T Professor of Engineering and Applied Sciences. Among his activities at the university are a complete revision of the undergraduate computer science curriculum, research on computer architecture and computer security, and a project to enable humanities scholars to exploit information technology. Dr. Wulf has had a distinguished professional career that includes serving as assistant director of the National Science Foundation; chair and chief executive officer of Tartan Laboratories Inc., Pittsburgh; and professor of computer science at Carnegie Mellon University. He is the author of more than 80 papers and technical reports, has written three books, and holds two U.S. patents.

Appendix C

Corporate Benchmarks

In preparation for the workshop on Best Practices in Managing Diversity, the NAE Committee on Diversity in the Engineering Workforce compiled a list of 71 companies that employ significant numbers of engineers and have been recognized for their handling of diversity issues in the workplace. All of these organizations either appeared in the *Fortune* list of the top 50 companies for minorities or the *Working Woman* list of the best companies for women, have been recognized by Catalyst or the Women in Engineering Programs and Advocates Network for their diversity programs, are represented on the Board of Directors of the National Action Council for Minorities in Engineering, or were recommended by a member of the committee.

Two UCLA students, Jill Bernardy and Jennifer Sommers, under the direction of committee member David Porter, collected available information about these companies' diversity management practices as benchmarks for the workshop participants. The information was collected from the World Wide Web and analyst's reports, supplemented by phone calls to the companies. Follow up phone calls were made by Professor Porter's research assistant Gina Kong to verify the accuracy of the information for each company. This data provided a starting point for discussions of diversity issues in the workplace.

The information was organized around the three stages of diversity program development described by David Thomas and Robin Ely in their article, "Making Differences Matter: A New Paradigm for Managing Diversity" (*Harvard Business Review* 74(5): 79–90). The three stages are described as (1) discrimination and fairness, (2) access and legitimacy, and (3) learning and effectiveness.

The discrimination and fairness paradigm focuses on equal opportunity, fair treatment, and compliance with legal requirements. The emphasis in this stage is on respecting differences and treating all employees equally but not on encouraging employees to apply their personal perspectives to their work. Programs at this stage seem to operate on the assumption that everyone is fundamentally the same.

The access and legitimacy paradigm recognizes and celebrates differences as a competitive advantage in reaching new markets. Companies operating under this paradigm are faced with increasing diversity among their customers, clients, or labor pool, and thus see diversity as a business opportunity or threat. This approach can lead to employees being pigeonholed into career paths that serve a particular market but do not lead to advancement in the company's mainstream operations.

Companies that go beyond the access and legitimacy paradigm to the learning and effectiveness paradigm begin to connect diversity to work perspectives. Companies with programs at the learning and effectiveness stage incorporate diverse outlooks of their employees into their work and take advantage of employees' different points of view to rethink and redesign their products and processes. More than just treating everyone fairly and celebrating differences, organizations operating under the learning and effectiveness paradigm are able to apply diverse perspectives to their fundamental business operations and improve performance.

The accompanying tables presents the diversity practices of the 71 companies classified into Thomas and Ely's three stages of diversity program development.

Table C-1

Table C-1 Best Practices Corporate Benchmark

	3M	Abbott Labs	Accenture	American Electric Power	Amgen	AOL	Applied Materials	AT&T	Baxter Healthcare	BE&K	Bechtel	Bell Atlantic	Bell South	Boeing	Bristol Meyers Squibb	British Petroleum	CH2MHill	Chevron	CISCO	Colgate Palmolive	Con Ed	Corning	Cummins Engine	Dow	DTE Energy
A-1 Regulatory Compliance - EEO, Affirmative Action		●		●	●	●	●		●		●	●		●	●	●		●			●			●	●
A-2 Successful at recruiting a diverse workforce (gender, race); Maintains an activity calendar of diversity recruiting events	●	●	●			●	●		●	●	●	●			●			●		●	●	●		●	
A-3 Race & Gender Awareness Program												●						●						●	
A-4 Benchmark within or outside of industry		●			●		●		●	●		●			●	●		●			●			●	●
A-5 Individual Growth-Horizontal: Access to experiences that yield personal and professional growth	●	●	●		●						●	●			●										
A-6 Individual Growth-Vertical: Equal representation among groups at all levels evident in business, functions, geographic areas															●										
A-7 Mentoring Program		●											●		●			●			●				
A-8 Diversity Mission or Value Statement		●								●		●		●	●	●								●	●
A-9 Training on cultural awareness/diversity		●		●					●	●		●	●	●	●	●		●					●	●	●
A-10 Sexual harassment training		●										●			●	●								●	●
A-11 Employee handbook refers to diversity issues; may require employee signatures															●										
B-1 Internal leadership committee (i.e., Executive Council on Diversity)		●							●									●							●
B-2 Educational partnerships and/or scholarships (i.e. to encourage women & minorities to study math & science)	●	●				●	●		●						●	●		●	●	●	●	●			
B-3 Internships for women and minorities		●							●									●	●		●	●			

153

B-4	Community outreach
B-5	Focus on **competence** based credentials rather than past experience
B-6	**Teambuilding training**
B-7	**Encourage & support partnerships** with minority affinity or network groups
B-8	**Celebrate multicultural events**
B-9	**Diverse Supplier Program**
B-10	Company **groups/clubs** for minority employees
C-1	Diversity is integral & valued **part of culture**
C-2	**Tool to measure** success of diversity programs
C-3	**Diverse Management Team**(s)/Board of Directors reflect diversity of current and future business environments
C-4	**Full-time diversity staff**
C-5	**Diversity objectives** are tied to corporate objectives
C-6	**Policies/ benefits** to support diverse needs (i.e. flex time, job-sharing, eldercare, daycare, seniority pay, etc.)
C-7	Administer regular **attitudinal surveys** (include diversity opinion surveys)
C-8	**Monitor and report** progress to staff
C-9	**Evaluate** business units' and managers' performance with regard to diversity; tie to bonus compensation
C-10	**Awards** for managers advancing companies diversity initiatives
C-11	Regular organization **newsletter** featuring all staff and highlighting internal & external diversity best practices
C-12	Formalized **succession/career development planning**
	Contacted by phone for verification

Table C-1 continued

	Duke Power	Dupont	EDS	Eli Lilly	Exxon Mobil	Ford	General Electric	General Motors	Goodyear	Hewlett Packard	Hoechst Celanese	Hughes	IBM	Idaho Nat'l EE Lab	Intel	Kodak	KPMG	Kraft	Lawrence Livermore Nat'l Lab	Lockheed Martin	Los Alamos Nat'l Lab	Lucent	Merck	Microsoft	Motorola
A-1 Regulatory Compliance - EEO, Affirmative Action	●	●	●		●	●	●	●		●		●	●	●				●	●	●					
A-2 Successful at recruiting a diverse workforce (gender, race); Maintains an activity calendar of diversity recruiting events	●	●		●	●		●	●		●			●		●			●	●			●	●	●	
A-3 Race & Gender Awareness Program	●	●				●	●	●				●							●						
A-4 Benchmark within or outside of industry	●		●		●	●	●	●	●				●					●		●					
A-5 Individual Growth-Horizontal: Access to experiences that yield personal and professional growth					●	●	●	●		●			●												
A-6 Individual Growth-Vertical: Equal representation among groups at all levels evident in business, functions, geographic areas								●	●	●										●		●	●		
A-7 Mentoring Program		●		●	●	●	●	●		●			●	●	●	●		●	●		●			●	
A-8 Diversity Mission or Value Statement					●		●	●																	
A-9 Training on cultural awareness/diversity		●	●	●	●	●	●	●		●										●					
A-10 Sexual harassment training					●	●		●																	
A-11 Employee handbook refers to diversity issues; may require employee signatures																									
B-1 Internal leadership committee (i.e., Executive Council on Diversity)	●	●	●		●		●	●		●			●		●	●		●	●	●	●				
B-2 Educational partnerships and/or scholarships (i.e. to encourage women & minorities to study math & science)						●	●			●												●	●		●
B-3 Internships for women and minorities	●	●	●		●	●	●			●			●	●	●			●				●	●	●	●

155

B-4	Community outreach
B-5	Focus on **competence** based credentials rather than past experience
B-6	**Teambuilding** training
B-7	**Encourage & support partnerships** with minority affinity or **network groups**
B-8	Celebrate **multicultural events**
B-9	**Diverse Supplier Program**
B-10	Company **groups/clubs** for minority employees
C-1	Diversity is integral & valued part of culture
C-2	**Tool to measure** success of diversity programs
C-3	**Diverse Management Team**(s)/Board of Directors reflect diversity of current and future business environments
C-4	**Full-time diversity staff**
C-5	**Diversity objectives** are tied to corporate objectives
C-6	**Policies/ benefits** to support diverse needs (i.e. flex time, job-sharing, eldercare, daycare, seniority pay, etc.)
C-7	Administer regular **attitudinal surveys** (include diversity opinion surveys)
C-8	**Monitor and report** progress to staff
C-9	**Evaluate** business units' and managers' performance with regard to diversity; tie to bonus compensation
C-10	**Awards** for managers advancing companies diversity initiatives
C-11	Regular organization **newsletter** featuring all staff and highlighting internal & external diversity best practices
C-12	Formalized **succession/career development planning**
	Contacted by phone for verification

Table C-1 continued

	Oak Ridge Nat'l Lab	Parsons Brinkerhof	PG&E	Proctor & Gamble	Public Service of NM	Qualcomm	Raytheon	Sandia Nat'l Lab	SBC	Schering-Plough	Sempra Energy	Shell	Silicon Graphics	Sony	Southern Cal Edison	Southern Company	Sun Microsystems	Texaco	Texas Instruments	Texas Utilities	US West/Quest	World Com	Xerox
A-1 Regulatory Compliance - EEO, Affirmative Action	•					•			•		•			•		•	•	•					
A-2 Successful at recruiting a diverse workforce (gender, race); Maintains an activity calendar of diversity recruiting events		•									•						•					•	•
A-3 Race & Gender Awareness Program									•		•		•	•		•		•					
A-4 Benchmark within or outside of industry				•							•		•						•				
A-5 Individual Growth-Horizontal: Access to experiences that yield personal and professional growth	•	•									•				•				•				•
A-6 Individual Growth-Vertical: Equal representation among groups at all levels evident in business, functions, geographic areas											•				•				•				•
A-7 Mentoring Program	•	•				•		•	•						•				•	•			
A-8 Diversity Mission or Value Statement	•											•							•				
A-9 Training on cultural awareness/diversity	•					•	•	•				•			•		•		•	•			
A-10 Sexual harassment training						•											•		•				
A-11 Employee handbook refers to diversity issues; may require employee signatures																							
B-1 Internal leadership committee (i.e., Executive Council on Diversity)								•				•			•					•			
B-2 Educational partnerships and/or scholarships (i.e. to encourage women & minorities to study math & science)	•			•	•	•					•		•										•
B-3 Internships for women and minorities											•				•					•			

B-4	**Community outreach**		•		•		•		•	•	
B-5	Focus on **competence** based credentials rather than past experience						•			•	
B-6	**Teambuilding training**			•							
B-7	**Encourage & support partnerships** with minority affinity or network groups		•	•		•			•		•
B-8	**Celebrate multicultural events**		•	•		• •					
B-9	**Diverse Supplier Program**		• •			•					
B-10	Company **groups/clubs** for minority employees	•				•			•	• •	
C-1	**Diversity is integral & valued part of culture**			•				•		•	
C-2	**Tool to measure** success of diversity programs				•						
C-3	**Diverse Management Team**(s)/Board of Directors reflect diversity of current and future business environments					•		•			
C-4	**Full-time diversity staff**	•				•		• •			
C-5	**Diversity objectives** are tied to corporate objectives	• •						•			
C-6	**Policies/ benefits** to support diverse needs (i.e. flex time, job-sharing, eldercare, daycare, seniority pay, etc.)						•				
C-7	Administer regular **attitudinal surveys** (include diversity opinion surveys)		•								
C-8	**Monitor and report** progress to staff						•				
C-9	**Evaluate** business units and managers' performance with regard to diversity; tie to bonus compensation		•	•					•		
C-10	**Awards** for managers advancing companies diversity initiatives	•									
C-11	Regular organization **newsletter** featuring all staff and highlighting internal & external diversity best practices									•	
C-12	Formalized **succession/career development planning**		•								
	Contacted by phone for verification	No	No	No	No	No	Yes	No	Yes	No	No

157